Lecture Notes in Statistics

Edited by D. Brillinger, S. Fienberg, J. Gani, J. Hartigan, and K. Krickeberg

32

Generalized Linear Models

Proceedings of the GLIM 85 Conference, held in Lancaster, UK, Sept. 16–19, 1985

Edited by R. Gilchrist, B. Francis and J. Whittaker

Springer-Verlag
Berlin Heidelberg New York Tokyo

Editors

Robert Gilchrist
Brian Francis
Joe Whittaker
Department of Mathematics, Statistics and Computing
The Polytechnic of North London
Holloway Road, London N7 8DB, UK

Mathematics Subject Classification (1985):

ISBN 3-540-96224-7 Springer-Verlag Berlin Heidelberg New York Tokyo
ISBN 0-387-96224-7 Springer-Verlag New York Heidelberg Berlin Tokyo

Printing and binding: Beltz Offsetdruck, Hemsbach/Bergstr.
2147/3140-543210

PREFACE

This volume consists of the published proceedings of the GLIM 85 Conference, held at Lancaster University, UK, from 16-19 September 1985.

This is the second of such proceedings, the first of which was published as No 14 of the Springer-Verlag Lecture Notes in Statistics (Gilchrist,ed,1982). Since the 1982 conference there has been a modest update of the GLIM system, called GLIM 3.77. This incorporates some minor but pleasant enhancements and these are outlined in these proceedings by Payne and Webb.

With the completion of GLIM 3.77, future developments of the GLIM system are again under active review. Aitkin surveys possible directions for GLIM. One somewhat different avenue for analysing generalized linear models is provided by the GENSTAT system; Lane and Payne discuss the new interactive facilities provided by version 5 of GENSTAT.

On the theory side, Nelder extends the concept and use of quasi-likelihood, giving useful forms of variance function and a method of introducing a random element into the linear predictor. Longford discusses one approach to the analysis of clustered observations (subjects within groups). Green and Yandell introduce 'semi-parametric modelling', allowing a compromise between parametric and non-parametric modelling. They modify the linear predictor by the addition of a (smooth) curve, and estimate parameters by maximising a penalised log-likelihood. Hastie and Tibshirani introduce generalized additive models, introducing a linear predictor of the form $\eta = \alpha + \Sigma f_j(x_j)$, with the f_j estimated from the data by a weighted average of neighbouring observations.

The theory of structural equation models, as for example considered in LISREL, is discussed by Arminger and Küsters in the context of categorical data. Similarly, Palmgren and Ekholm provide an example of a GLIM program to apply latent class analysis to data on four dichotomised variables, with respondents classified as high or low IQ.

Multidimensional tables continue to provide a rich field for the application of glm's. Havránek and Pokorný illustrate the GUHA approach; Whittaker considers the effect of perturbing contingency tables, with particular reference to the effect upon the useful class of graphical models. The question of degrees of freedom in sparse tables where there are marginal zeros was discussed by Aston and Wilson (1984), in reply to Brown and Fuchs (1983). In these proceedings, Haslett supports the view that weighted least squares can be misleading. (Readers may wish to compare this with Aston and Wilson's suggestion that there is no real difficulty as the standard errors clearly demonstrate that the parameter estimates are not correct. The fitted values nevertheless are correct).

In recent years, glm's have been seen as an effective means of analysing ordinal data. In these proceedings, the problem of estimating interobserver variation for ordinal rating scales is considered by Jørgensen, with an example on the diagnosis of arthritis. Other authors deal with a variety of theoretical and practical problems. Cordeiro extends his work on expected deviance to the class of generalized linear models introduced by Jørgensen (1983). Diggle shows how to use a glm to compare estimated spectral densities. Stukel proposes a form of link function which generalises the logistic formulation; she illustrates this by application to the well-known beetle mortality data.

There is an introductory paper by Gilchrist, stressing the role played by GLIM in the development of glm's and illustrating how diagnostic procedures arise naturally from their use. All other papers are arranged in alphabetical order, by author.

Thanks are due to the following for their help in refereeing contributed papers:

M Aitkin, R J Baker, M R B Clarke, B Francis, B Jorgensen, M Green, C Payne, R Rigby, A Scallan, C Sinclair, R Thompson, J Whittaker.

References

Aston, C E and Wilson, S R (1984). Comment on M B Brown and C Fuchs, 'On maximum likelihood estimation in sparse contingency tables'. *Computational Statistcs and Data Analysis,* 2,71-77.

Brown, M B and Fuchs, C (1983). On maximum likelihood estimation in sparse contingency tables. *Computational Statistics and Data Analysis,*1, 3-15.

Gilchrist, R,ed (1982). *GLIM82: Proceedings of the International Conference on Generalised Linear Models*. New York:Springer-Verlag.

Jørgensen,B (1983). Maximum likelihood estimation and large sample inference for generalized linear and non-linear regression models. *Biometrika,*70,19-28.

RG
Department of Mathematics,
Statistics and Computing,
Polytechnic of North London, UK.

GLIM 85: CONTENTS

INTRODUCTION: GLIM AND GENERALIZED LINEAR MODELS

By ROBERT GILCHRIST

Polytechnic of North London,UK.

SUMMARY

This paper discusses the key role of the GLIM statistical package in the development of generalized linear models. With the release of GLIM 3.77, it is an appropriate time to record the key role of members of the GLIM Working Party in developing GLIM. The use of glm's as a technique for data analysis is illustrated by reference to arising diagnostic procedures which arise naturally from the glm approach.

Keywords: GLIM; Generalized linear models; diagnostics; residuals; partial residuals; estimation; mean-variance; link.

1. GLIM and the development of generalized linear modelling.

An interest in generalized linear models is richly rewarded. Not only does it bring a wealth of interesting theoretical problems but it also encourages an ease of data analysis sadly lacking from traditional statistics. It is my view that the role of GLIM cannot be stressed sufficiently here; the availability of a robust package, with legitimate results and a convenient model definition syntax, has revolutionised statistical analysis. There have been many criticisms of the GLIM system: the manual is poor, it was never designed for use by non-statisticians, it is difficult to get started with, the weighted least squares algorithm is too restrictive for many problems, APL provides a more general and convenient tool, and so on.

There may be some validity in these arguments; indeed, my original interest in GLIM was stimulated by my intuition that it has something valuable to offer, if only I could get it to work! The greatest step forward was being able to use GLIM interactively. And the manual has improved. Morever, it can be argued that the very terseness of the GLIM syntax marked a significant trend for interactive software. GLIM paved the way by having interactive help facilities, even if they have sometimes not been suffficiently explicit for the novice. Getting started is still a problem; 'Primers' can help here, and the new manual now contains an elementary introduction, although there is still no real substitute for the helping hand of an experienced instructor. The inefficiency of weighted least squares in, for example, analysing a multiway contingency table is undeniable; However, computational time and storage are usually not real problems in this age of increasingly cheap hardware. Thus, for example, it is my contention that for the user who wants an occasional analysis of a multiway table, the glm approach has few drawbacks. On the other hand, the specialist with large tables will naturally look for a more efficient algorithm. Nevertheless, an added bonus of the glm approach is the insight provided by imbedding a problem in a wider context; this in itself encourages a more critical approach to data analysis and leads naturally to, for example,thinking about the need for diagnostic procedures. (see section 2, below).

It is, in my view, clear that the blossoming of analyses related to generalized linear modelling is directly due to the availibility of GLIM. The average non-statistician and non-computing specialist cannot effectively handle both the theory of weighted least squares and its programming in a high level language. There are of course experts who can do this; see e.g Green (1984). And it is true that, if you can manage both the statistical theory and the intricacies of APL, then GLIM is sometimes not the best vehicle for analysing as given problem, or even classes of problems. But we would surely never have arrived at this position without the bedrock of GLIM, on which many of us have developed an understanding of the power of the original formulation of glm's by Nelder and Wedderburn (1972). We owe the late Robert Wedderburn a great debt for this contribution to modern statistics.

At the same time, the role of John Nelder cannot be stressed sufficiently; his guiding hand has steered GLIM into existence, not least by programming many of the routines himself. It is appropriate that John Nelder's contribution to statistics has been formally recognised by his recent election to the Presidency of the Royal Statistical Society. At the same time, his research activities continue, both in statistics and computing. One aspect of this is reflected in these proceedings, with extensions of Wedderburn's (1974) work on quasi-likelihood. John Nelder's current project on intelligent software will doubtless bring similar advances. Such 'expert sytems' are still in their infancy, but they offer great potential; it would be all too perilous to leave their development to non-statisticians. Statisticians have tended to be all too insular in the past; a prime example is the divergence of Operations Research from statistics. Such an attitude can only lead to the downfall of the statistics profession. (Readers interested in a discussion of how an expert system can be devised for regression analysis might like to refer to Pregibon and Gale, 1984).

One of the main advantages of GLIM has been its statistical and numerical reliability. In this respect, the work of Michael Clarke has received insufficient attention; his programming of the computational algorithm of GLIM has been one of the most important strengths of the package. Indeed, one of the unfortunate aspects of the demise of the Prism system (Baker, 1982) is the non-appearance of Michael Clarke's implementation of the Given's algorithm within GLIM. This orthogonal decomposition approach (see also Clarke, 1981) is as accurate in single precision as is Gauss-Jordon in double precision. Moreover, Michael Clarke had applied much effort to obtain values of tolerances, etc, which led to identical results for the two approaches.

The Prism system is now dead, although some of its facilities are included in the new release of GLIM; see Payne and Webb in this volume. In particular, Michael Green's TABULATE directive gives GLIM 3.77 many of the facilities of the popular teaching package MINITAB (Ryan, Joiner and Ryan, 1985).

Payne and Webb acknowledge the work of many contributors to GLIM 3.77. A team of statisticians and programmers is the only effective way to develop and maintain a statistical system (buyers of statistical software for micros beware; see Gilchrist, 1984a). Payne and Webb rightly record our debt to Bob Baker who has played the major role in the programming of both GLIM 3 and GLIM 3.77. (unfortunately, Bob Baker will not be able to carry out this task for the next release, although he continues his active interest in GLIM).

Developing a major statistical package is a fairly big task. Something of the order of 50-100k lines of code might be needed; the investment in time (and money) has been substantial; a new version may require funding of the order of hundreds of thousands of pounds sterling. Many of the difficult questions which bedevilled Prism are only now being resolved; e.g. what statistical facilities should be made available, what numerical facilities, what form(s) of user interface? Aitkin's contribution to this volume provides some suggestions for (mainly) statistical developments; this conference allows this debate to be brought to a wider audience.

2. Diagnostics: some comments

As mentioned previously, one of the attractions of glm's is that an analysis is naturally imbedded in a wider framework; this leads directly to ideas for diagnostic procedures. This section outlines some approaches which may be useful; the general point being stressed is that these ideas apply to traditional (non-glm) analyses. By viewing a statistical technique as a glm, we are naturally led to diagnostics for traditional techniques.

2.1. *Residuals*

Residuals play a key part in diagnostic procedures but there is no firm decision as to which type of residual should be used. The traditional 'standardised' residuals are defined as sres = $(y-\mu)/\sigma$, where μ=E[Y] and σ^2=Var[Y] are estimated from the fitted model. (GLIM effectively uses this definition, although it does not estimate any unknown scale). It seems desirable to adjust the sres to allow for the correlation beyween y and μ. This leads to ares = sres/$\sqrt{(1-h)}$, where h is easily calculated in GLIM as %WT*%VL/%SC (Gilchrist, 1984b). The 'deviance residual' is also informative; this is dres = $\sqrt{d}$ sgn(y-μ), where d is the deviance contribution of the observation y. Normal and half Normal plots can be used to assess Normality of data; when these are applied to non-Normal data a transformation can be applied before plotting. For example, plotting $\sqrt{y}-\sqrt{\mu}$ for Poisson data. A general alternative is the 'Anscombe residual', defined by replacing y by A(y). (See, e.g., McCullagh and Nelder,1983). The aforementioned adjusted residuals have the nice property of providing the score statistic for testing that a particular observation is an outlier (Pregibon, 1982). Moreover, they form the basis of a method for detecting whether a particular observation is unduly influential. This is done by calculating c^2 = ares^2*h/(1-h) and noting for which points this is 'large'. Such points often stand out well, although an envelope can be constructed. An alternative is to use h itself and note for which points this is large. Here, 'large' is often taken to be 2p/n or 3p/n, when p parameters have been fitted to n observations, although this rule appears not to have any firm foundation for non-Normal data.

Unfortunately, detecting non-Normality from Normal plots can be tricky with few observations. Moreover, non-linearity of such plots can be due to a variety of different causes; it is therefore useful to look at ways of distinguishing different forms of model inadequacy.

2.2. *The mean/linear predictor relation.*

Pregibon (1980) introduced a method of testing the adequacy of the link function of a glm by imbedding the link in a parametric family with extra unknown parameters. Expanding the link as a Taylor's series in these parameters and fitting extra variates gives a test of the original link. For example, with the Box-Cox link defined by g(μ) = (μ**b-1)/b, the log link is tested by fitting the extra variate $\log^2(\mu)/2$. The change in deviance gives a measure of the inadequacy of the original link. One problem with this method is that the assumed form of link may be some way from the 'true' value, so the Taylor's series expansion may not be sufficiently accurate. An alternative method, preferred by this author, is to plot the deviance for a set of values of the unknown parameter(s), in this case b. This the allows the maximum likelihood estimates and interval estimates to be found, by eye. Scallan, Gilchrist and Green (1984), discuss a stable 2-stage procedure for the fitting of the Box-Cox link with unknown exponent. This procedure is not necessary if all we require is a fairly good idea of an interval estimate of the unknown link parameter(s), as Scallan, Gilchrist and Green indicate that a 100(1-α)% interval estimate for b is given, for the case of error distribution with known scale, by

$$[b: D(b) \leq D(b) + \chi^2_{1,\alpha}],$$

where D(.) is the usual deviance expression and b is the observed maximum likelihood estimate of b. In addition, it should be noted that the Box-Cox link gives the same deviance as the power link, provided a constant term is fitted. Thus, for example with Poisson distributed data, an interval estimate for b is readily found by fitting the power link (available as $LINK E %B] and plotting deviances against b.

The case with unknown scale in the error distribution is treated similarly, giving interval estimates of either

$$[b: D(b) \leq D(b)\{ 1 + \chi^2_{1,\alpha}/(n-p-1)\}],$$

or

$$[b: D(b) \leq D(b) \exp\{\chi^2_{1,\alpha}/n\}],$$

depending on whether it is preferred to estimate the unknown scale at the maximum likelihood estimate of b or at its current estimate.

2.3 *The mean/variance relation.*

A test of the adequacy of the mean/variance relation can be developed by assuming the variance function V(μ) depends on an unknown parameter θ; a useful family is var(y) = $\phi.\mu^\theta$, where ϕ is the scale parameter. Such a model can be easily fitted once the deviance is determined. It was shown in Gilchrist et al (1982) that, for the exponential family, the 'exponent term' required is $\frac{1}{2}D(y,\mu,\theta)$, where

$$D(y,\mu,\theta)=-2y(\mu^{**}(1-\theta)-y^{**}(1-\theta))/(1-\theta)+2(\mu^{**}(2-\theta)-y^{**}(2-\theta))/(2-\theta)$$

with the Poisson and Gamma forms being used for $\theta=1,2$ respectively. Use of the 'constant term' in the deviance is essential when fitting to allow for the different magnitude of D for various θ. As Nelder and Pregibon (1985) show, the intuitively sensible 'quasi-likelihood' given by,

$$Q(y,\mu,\theta)=-\tfrac{1}{2}\ln(2\pi\phi y^{**}\theta)-\tfrac{1}{2}D(y,\mu,\theta)/\phi,$$

is in fact the saddlepoint approximation for exponential families. Use of this technique leads, for example, to joint interval estimates for (b,θ).

2.4. *Partial residual plots.*

We have so far described glm based diagnostics which may help in traditional analyses; we should also remark on some other new developments in diagnostic procedures for glm's, based upon traditional techniques.

So-called partial residual plots are quite widely used in conventional regression (see e.g Velleman and Welsch, 1981) to diagnose non-linear relationships between response and explanatory variates and to attempt to transform an explanatory variate to give a linear relation to the response. Landwehr, Pregibon and Shoemaker (1984) propose a similar procedure for logistic regression. The partial residuals are plotted against a single explanatory variate, with a linear relationship giving a linear plot. However, because of the use of a dichotomous response, smoothing of the plot is required, using both robust and non-robust methods. A practical application of such techniques is given by Wrigley and Dunn (1985).

3. Conclusion.

Generalized linear models provide a wealth of interesting theoretical problems as well as being a powerful data-analytic tool. There is active work on extending the theoretical concepts ; see for example, McCullagh (1983), Jorgensen (1983, 1984). In addition, there is extensive work on new applications. This conference demonstrates the increasing international flavour of these developments and indicates some of the potential of the approach. The future clearly holds much in store for glm's; we owe much to its originators.

References.

Baker, R J (1982). Prism,an overview. In: *GLIM 82: Proceedings of the international conference on generalised linear models*. (Gilchrist,R.ed). New York:Springer-Verlag.

Clarke, M R B (1981). Algorithm AS163. A Givens algorithm for moving from one linear model to another without going back to the data. *Appl. Statist.*, 30,198-203.

Gilchrist, R.(1984a) Contribution to Discussion in Green,P.(1984) op cit.

Gilchrist, R.(1984b). The Polytechnics' micro evaluation project. *The Professional Statistician*,7,17-19.

Gilchrist, R. and Scallan, A.(1984) Link functions in generalised linear models. *COMPSTAT* 84, 203-208, Wien: Physica Verlag.

Gilchrist, R., Green, M. and Scallan, A.(1982) Testing the mean-variance relationship and the link function in generalised linear models. *Supplement to COMPSTAT* 82.

Green, P.(1984) Iteratively reweighted least squares for maximum likelihood estimation and some robust and resistant alternatives. *J.R.Statist.Soc.* B, 46,149-92.

Jorgensen, B. (1983) Maximum likelihood estimation and large sample inference for generalized linear and non-linear regression models. *Biometrika*, 70, 19-28

Jorgensen, B. (1984). The delta algorithm and GLIM. *International Statistical Review*,52,283-300.

Landwehr, J M, Pregibon D and Shoemaker, A C (1984). Graphical methods for assessing logistic regression models. *Journal Amer Statist Ass*,79,61-71.

McCullagh, P. (1983) Quasi-likelihood functions. *Ann. Statist.*, 11, 59-67.

McCullagh, P and Nelder, J A.(1983). *Generalized linear models*. London: Chapman and Hall.

Nelder, J.A. and Pregibon, D. (1985) Quasi-likelihood functions and data analysis. *Bell Laboratories Technical Memorandum*.

Nelder, J.A. and Wedderburn, R.W.M.(1972) Generalised linear models. *J.R.Statist.Soc.*, A, 135, 370-84.

Pregibon, D. (1980) Goodness of link tests for generalised linear models. *Appl. Statist.*, 29, 15-24.

Pregibon, D. (1982) Score tests in GLIM with applications.In: GLIM 82: *Proceedings of the International Conference on Generalised Linear Models* (Gilchrist,R.ed). New York:Springer-Verlag.

Pregibon, D and Gale,W. (1984). REX: an expert system for regression analysis. *COMPSTAT 84*. 242-248. Wien:Physica Verlag.

Ryan B F, Joiner,B L and Ryan, T A.(1985) *Minitab Handbook*. 2nd Edition. Boston:Duxbury.

Scallan, A, Gilchrist,R and Green,M.(1984) Fitting parametric link functions in generalised linear models. *Comp.Statist and Data Anal.*,2, 1, pp 37-49.

Velleman, P F and Welsch R E. (1981) Efficient computing of regression diagnostics. *The American Statistician*, 35,234-41.

Wedderburn, R.W.M. (1974) Quasi-likelihood functions, generalized linear models and the Gauss-Newton method, *Biometrika, 61*, 439-47.

Wrigley,N and Dunn, R.(1985) Graphical diagnostics for logistic oil exploration models. Univ Bristol: Dept Geography Report.

GLIM4 - DIRECTIONS FOR DEVELOPMENT

Murray Aitkin
Centre for Applied Statistics, University of Lancaster

1. SUMMARY

A personal view of GLIM development is described. It is proposed that GLIM should be substantially developed by the expansion of its range of standard error distributions, which should include general mixture distributions fitted by nonparametric maximum likelihood, by the enhancement of the programming language to include matrix and array operations and eigenvalue-eigenvector decompositions, and by the enhancement of its present graphics facilities. These developments could be funded initially by research grant support from appropriate bodies; other sources of development capital should be actively explored. During this period an appropriate organisation should be established to market and support GLIM and to continue its development; this would require generation of income by a change from outright sale to annual leasing of GLIM. International sales could be decentralised by the appointment of national distributors with responsibility for local courses, supported by a proportion of income from national sales.

2. BACKGROUND

Within a relatively short period, GLIM has established itself internationally as a small but powerful statistical package with a unique modelling philosophy. Nelder and Wedderburn's fundamental paper (1972), translated into an efficient computing system, has fundamentally changed our approach to statistical theory and data analysis. The aims of the developers, as expressed in the name GLIM, were relatively modest : to fit interactively regression models in the one-parameter exponential family, accommodating also the two-parameter normal and gamma distributions. Nevertheless by the time GLIM3 was released in 1978 it was quickly appreciated by users that the calculating language and macro facilities allowed a vastly greater range of models to be fitted than the exponential family models provided as standard in the $ERROR options. Macros have been developed for fitting the negative binomial and the censored exponential, Weibull, extreme value and logistic distributions. Macros based on various EM algorithms have been able to fit a remarkable variety of 'missing data' models including the censored and grouped normal, latent class models and finite mixture distributions.

With this appreciation of the great possibilities of GLIM came an increasing appreciation of its limitations. The much wider application of iteratively reweighted least squares outside the exponential family was pointed out by Jorgensen (1984) and Green (1984). The Weibull and extreme value distributions, for example, can be fitted by maximum likelihood using IRLS, including the estimation of the scale parameter, and censoring can be handled easily if the iterative weights can themselves be different for the censored and uncensored observations. This approach is quite general, and does not require the special Poisson feature of the likelihood which allows the macro approach by Aitkin and Clayton (1980) and Aitkin and Francis (1980).

Even within the GLIM exponential family models there are limitations. The shape parameter of the gamma distribution is not estimated in GLIM; the 'mean deviance' approach is an approximation of unknown accuracy. Survival modelling using the gamma distribution is not possible with censored observations which would require the incomplete gamma function for the likelihood of the censored observations.

The main limitation of the calculating language is the inability to access the SSP matrix and to perform matrix operations. Both facilities are essential in nearly all EM algorithm problems, where the standard errors based on the 'complete data' information matrix printed out by GLIM in the M-step are incorrect - in the non-conservative direction - though the parameter estimates themselves are correct. Currently there is no way around this problem except to omit each variable and compute the change in the true deviance. Standard errors produced by the Weibull macro are underestimated for the same reason, though it is possible to correct this by the device of an extra observation, due to Roger and Peacock (1983).

A minor but still important point is the slow running of macros, particularly when these are nested. The re-interpretation of the macro in each iteration is very wasteful of CPU time.

3. THE RELATION OF GLIM TO OTHER STATISTICAL PACKAGES

After six years of success, and increasing sales and use of GLIM3 worldwide, GLIM is assured of a continued future for some years, even in its present form. There is a case for continuing it in this form, as a powerful and useful adjunct to the main scientific statistical packages: GENSTAT, BMDP and SAS (SPSS is in a different category). GLIM is too restricted in its programming language and its range of facilities to be a real competitor to any of these systems, but it provides powerful and simple model fitting within the class of distributions it was designed to handle. In its present form it could be successfully marketed for micro-computers with adequate main memory and disc store. Since the model fitting routines in GLIM are also provided in GENSTAT (apart from the $OWN routines), a simple answer to the complaint of lack of facilities in GLIM is to refer the user to GENSTAT.

This is not an answer which finds favour in the US market. GENSTAT is hardly used in the US, where SAS is the most widely used statistical package, partly because of its excellent graphics routines. The inability to program matrix routines in GLIM means that SAS must be used for such routines; this limits in turn the attractiveness of GLIM in the US market. Since it takes a major effort to learn <u>any</u> programming language, why learn two?

I believe that GLIM is now at a critical point. The enthusiasm and very hard voluntary work of the developers have brought GLIM to a respected, if minor, position in the international statistical package market. Sustaining even the present level of support for GLIM3.77 is placing a severe strain on the GLIM Working Party, which cannot continue much longer as a group of volunteers without adequate financial backing.

As the new Chairman of the Working Party in succession to John Nelder, I naturally want GLIM to go forward, and to achieve the international success that it deserves on a much larger scale. How is this to be done?

4. A STRATEGY FOR GLIM DEVELOPMENT

Assuming that GLIM is to be further developed, a strategy for its development requires a definition of GLIM's future role in the statistical package market. The strategy followed here is that GLIM should remain a maximum likelihood statistical modelling package, but with a much broader range of standard probability distributions, an expanded calculating language including full matrix and array operations and singular value decompositions, and expanded graphics facilities. GLIM should be recognized and marketed (particularly in the US) as a statistical package with an advanced theoretical and computational design, able to fit a very broad range of statistical models and to give adequate graphical representations of the fitted models and data. It would then be a serious competitor for SAS and BMDP in most, if not all, scientific applications outside time series. A version for large micros would be particularly attractive in American and European computing environments where mainframe time costs real money.

Such a strategy does not imply the incorporation of the Wilkinson (1970) algorithm for generally balanced designs as used in GENSTAT, which was proposed for PRISM. Although it is naturally an advantage to be able to produce efficiently the ANOVA table for a balanced design, this ability does not extend to non-normal models, and general unbalanced variance component models cannot be handled efficiently by the Wilkinson algorithm. The emphasis on fitting a sequence of specific models, which is the key to the understanding of the analysis of unbalanced cross-classifications (Aitkin 1978), is also lost in the use of the GENSTAT analysis. Most of the social and medical applications of GLIM, the fields in which it has had the greatest impact, do not have balanced multifactor designs, in any case.

The details of the proposed additions to GLIM are given in §5, and possible structures for its development, marketing and support are discussed in §6. I must stress that the views expressed here are personal, and not those of the Working Party or the RSS, which at the time of writing have not discussed them.

5. ADDITIONS TO GLIM

5.1 Error distributions for censored data

Green (1984) pointed out that for any location/scale parameter family IRLS can be used to fit a generalized linear model to the location parameter, and the scale parameter can be estimated at each iteration from a simple weighted sum of squares using the fitted linear predictor. Thus the exponential, Weibull and extreme value distributions can be fitted efficiently using IRLS when there is no censoring without recourse to the Poisson likelihood. The present difficulty of efficient fitting of censored survival data remains.

It is proposed that these distributions, and the piece-wise exponential, normal and gamma, be available with an option allowing arbitrary left- and/or right-censoring. The weights used in the IRLS fitting would be defined by an option variable taking values 1, 2 and 3 for uncensored, left- or right-censored observations. For example $ERROR N C would recognize the censoring indicator in C. Currently the censored normal can be fitted in GLIM using an EM algorithm (Aitkin 1981) but standard errors may be wildly mis-stated by the GLIM 'complete data' parameter estimate covariance matrix.

For the gamma distribution this will require the incomplete gamma function for the weights, and an efficient (ML) estimation of the scale parameter, which has currently to be assumed to be known, or estimated from the mean deviance.

5.2 <u>Multinomial distributions</u>

The current method of fitting the multinomial distribution for multi-category responses uses the multinomial/Poisson relation. This is adequate for small contingency tables but cannot handle even a moderate-sized set of data with continuous explanatory variables and a polytomous response. A proper multinomial logit/probit model fitting routine is required. This could be achieved in several ways; one simple way would be to use the Cholesky decomposition of the non-diagonal covariance matrix of the category responses to convert the model to a new design matrix with independent responses, as suggested by Wilson and others in the discussion of Green (1984).

Ordered categorical responses also need to be incorporated into GLIM : the work by McCullagh (1980) and Anderson (1984) shows how straightforward this is in comparison to composite link function approaches.

5.3. <u>General mixture distributions</u>

Finite mixture distributions are becoming increasingly important in practical applications. Recent work by econometricians and American sociologists on the effect of omitting important variables from the model has shown that the estimation of regression parameters in logit and hazard function models can be seriously biased by the failure to model the 'extra variation' through a compound distribution. The difficulty of specifying a parametric form for the 'mixing' distribution of the omitted variable, or equivalently for the marginal compound distribution of the response variable, can be avoided completely by <u>estimating</u> the distribution of the mixing variable by nonparametric maximum likelihood. A discussion of economic and social applications of such models is given in the book by Diekmann and Mitter (1984), which contains numerous references to their analysis using GLIM.

Current work by Hinde and Wood at Exeter has established the feasibility of fitting arbitrary mixtures of exponential family distributions using GLIM macros and estimating the mixing distribution by nonparametric maximum likelihood as a discrete distribution on a finite number of mass-points. The computational method using the EM algorithm was described by Laird (1978). The importance of this result for GLIM development is that the <u>same</u> computational routines can be applied to <u>any</u> exponential family distribution, and to the Weibull/ extreme value family, with the additional estimation of the scale parameter for the normal, gamma and Weibull distributions.

There are three important and quite distinct applications of this approach :

i) Variance component models for two-level nested designs can be fitted for normal, binomial or Poisson distributions without assuming <u>any</u> parametric model for the random effect. This greatly extends the present limited range of such models. More than two levels of nesting can be accommodated if necessary, though the number of mass-points becomes appreciable.

ii) Models for overdispersion - that is, general compound distributions - can be fitted using an "extra variation" error term in the linear predictor for binomial, Poisson, gamma and Weibull models, without <u>any</u> parametric model for the error term.

This result is outstandingly important in the analysis of longitudinal data by semi-Markov models. The fitting of compound models to unemployment duration and risk is an essential part of the Lancaster/ Manchester team study of unemployment experience in Rochdale, which has received substantial ESRC support from the Economic Life Initiative of the Social Affairs Committee. Part of the project is the development of efficient procedures for fitting compound models by nonparametric ML. This could form part of the first stage of the proposed GLIM development outlined in this paper.

iii) Robust regression models can be fitted for the normal distribution with outlier contamination using an additional error term in the linear predictor with an arbitrary distribution which is estimated as a finite mixture. If there are no outliers, the mixing distribution will be estimated as a single component and the standard normal analysis will result. This approach is both more flexible and more theoretically justifiable than either standard "robust" methods or parametric methods based on fitting the t-distribution.

The current implementation of the EM approach as GLIM macros by Hinde and Wood requires the blocking of the data into a vector of length $n \times \ell$, where n is the sample size and ℓ the number of mass-points in the model. This is both slow and very wasteful of space. Efficient generation of the weighted SSP matrix is not difficult, and would work in the same way for any exponential family model, and for the Weibull. The implementation of this feature alone would put GLIM well ahead of <u>any</u> other statistical package. For the analysis of multi-stage nested designs with a normal response, many covariates and several levels of nesting, the mass-point method would be very slow, and the implementation of the fast Fisher scoring algorithm by Longford (1985) for normal models would be an important, though limited, complement to the general mixture routines.

5.4 Matrix and array operations

However far the current model-fitting features of GLIM are extended, there will always be complex models requiring special macros. A major deficiency of GLIM is its lack of a matrix calculus to produce the correct information matrix for non-standard models. This is most visible in EM algorithm applications but is also true in a range of other models, for example the Weibull. The abortive PRISM development would have provided general matrix and array operations : these should be incorporated (not necessarily in their PRISM form) in GLIM together with a singular value decomposition routine and Cholesky factorization. This would allow quite general non-diagonal covariance matrix structures to be analysed in simple macros; standard multivariate analysis procedures could also be developed as macros, though it is neither necessary nor appropriate to develop full MANOVA procedures.

5.5 Graphics

GLIM will not be able to compete effectively with other statistics packages (particularly SAS and MINITAB) without reasonable graphics facilities. These could be developed initially through the activation of the non-active routines in GLIM3.77. At a minimum curve plotting, scatterplots and contouring should be available by simple directives.

6. FINANCIAL SUPPORT, MARKETING AND DOCUMENTATION

This conference is not the appropriate forum for a discussion of finances, but a few comments can be made. Much of the difficulty with the present support of GLIM3 is that the package is sold outright very cheaply. The current income from GLIM sales could not support more than one person full-time. If long-term support and development of GLIM is to continue on an appropriate scale, it will be essential to change from outright sale to annual leasing of GLIM4. An effective development, support and marketing organisation can then be assured of the expanding income necessary to maintain and expand its activities.

We should not underestimate the costs and difficulties of international promotion. One way of increasing international visibility would be through the appointment of 'national distributors' in each country, who would be responsible for organising GLIM courses and demonstrations in that country with support from the GLIM organisation, and who would receive a proportion of the income from national sales.

In providing support, a major effort is required by the Working Party and the GLIM organisation in documentation. The GLIM3 Primer produced by Gilchrist and Green has been quite successful, and a new manual, primer and reference guide will be available for GLIM3.77, as well as the textbook 'Statistical Modelling in GLIM3' by Aitkin, Anderson, Francis and Hinde (1985). For GLIM4 a much more general series of monographs will be required on specific topics. The aim of such a series should be to cover the whole field of applications possible in GLIM4, with a particular eye on the US college course market, where there is very little published in computational statistics and statistical modelling. Adena and Wilson's (1982) monograph on case/control studies is an example of the possibilities.

CONCLUSION

GLIM is now at the crossroads. With proper development it can become a highly successful international advanced statistical modelling package. Development is expensive, but the principal difficulty is not attracting finance for the initial development, but ensuring the long-term future of GLIM with an active, stable and effective supporting organisation.

REFERENCES

Adena, M.A. and Wilson, S.R. (1982). Generalized Linear Models in Epidemiological Research ; Case Control Studies. Sydney : Instat Foundation.

Aitkin, M.A. (1978). The analysis of unbalanced cross-classifications (with Discussion). J.Roy.Statist.Soc.A. 141, 195-223.

Aitkin, M.A. (1981). A note on the regression analysis of censored data. Technometrics, 23, 161-163.

Aitkin, M.A., Anderson, D.A., Francis, B.J. and Hinde, J.P. (1985). Statistical Modelling in GLIM3. Oxford: University Press (to appear).

Aitkin, M.A. and Clayton, D.(1980). The fitting of exponential, Weibull and extreme value distributions to complex censored survivaldata using GLIM. Appl.Statist. 29, 156-163.

Aitkin, M.A. and Francis, B.J. (1980). A GLIM macro for fitting the exponential or Weibull distribution to censored survival data. GLIM Newsletter 3, 19-25.

Anderson, J.A. (1984). Regression and ordered categorical variables (with Discussion). J.Roy.Statist.Soc.B, 46, 1-30.

Diekmann, A. and Mitter, P. (1984). Stochastic Modelling of Social Processes. Orlando : Academic Press.

Green, P.J. (1984). Iteratively reweighted least squares for maximum likelihood estimation, and some robust and resistant alternatives (with Discussion). J.Roy.Statist.Soc.B, 46, 149-192.

Jorgensen, B. (1984). The delta algorithm and GLIM. Int.Statist.Rev.52.

Laird, N.M. (1978). Nonparametric maximum likelihood estimation of a mixing distribution. J.Amer.Statist.Assoc. 73, 805-811.

Longford, N.T. (1985). A fast scoring algorithm for maximum likelihood estimation in unbalanced mixed models with nested random effects. Submitted to J.Amer.Statist.Assoc.

McCullagh, P. (1980). Regression models for ordinal data (with Discussion). J.Roy.Statist.Soc.B, 42, 109-142.

Nelder, J.A. and Wedderburn, R.W.M. (1972). Generalized linear models. J.Roy.Statist.Soc.A. 135, 370-384.

Roger, J.H. and Peacock, S.D. (1983). Fitting the scale as a GLIM parameter for Weibull, extreme value, logistic and log-logistic regression models with censored data. GLIM Newsletter 6, 30-37.

Wilkinson, G.N. (1970). A general recursive procedure for analysis of variance. Biometrika, 57, 19-46.

SIMULTANEOUS EQUATION SYSTEMS WITH CATEGORICAL OBSERVED VARIABLES

Gerhard Arminger and Ulrich Küsters
Department of Economics, Bergische Universität Wuppertal
D-5600 Wuppertal, Federal Republic of Germany

SUMMARY

Various techniques to estimate structural equation models with metric latent variables measured indirectly by unordered categorical observed variables are proposed.
KEYWORDS: AMEMIYA'S PRINCIPLE, EM-ALGORITHM, GLIM, LATENT TRAITS, LISREL, MARGINAL LIKELIHOOD, MULTINOMIAL LOGIT, SIMULTANEOUS EQUATION SYSTEMS

1. MODEL FORMULATION

Econometricians and psychometricians increasingly use simultaneous equation models with latent variables. The LISREL approach extended by Muthén (1984) connects metric and ordinal manifest variables with metric latent variables. However, Muthén's model cannot deal with unordered categorical manifest variables. Bock (1972), on the other hand, has proposed a multinomial logit model for latent traits, combining categorical variables with a metric latent variable. His model does not contain explanatory variables and hence it cannot be used for structural analyses. To develop realistic models of economic and social behavior, a combination of the models above is necessary. For example, suppose that one wants to study the relationship between political interest, political tendency, and various exogenous variables such as income, education and sex. Political interest is typically measured on an ordinal scale and politically tendency is measured indirectly by unordered categorical variables such as party preference and opinion about the structure of society. In this instance the models above are insufficient. Hence, we propose techniques to estimate structural equation models with metric latent variables measured indirectly by unordered categorical observed variables.

For simplicity's sake, we consider a system with two equations:

$$\eta_1 = \beta_1\eta_2 + \underline{\gamma}_1^T\underline{x}_1 + \varepsilon_1, \qquad (1.1)$$

$$\eta_2 = \beta_2\eta_1 + \underline{\gamma}_2^T\underline{x}_2 + \varepsilon_2.$$

It is assumed that $(\varepsilon_1,\varepsilon_2)^T$ is distributed as $N_2(\underline{0},\underline{\Sigma})$, where $\underline{\Sigma}$ is positive definite, and the rank and order conditions for identification hold true (Schmidt, 1976). To ensure identification of the logits in the measurement model subsequently introduced the usual constant term in $\underline{x}_1$ is omitted. To obtain the reduced form, we write equation (1.1) in matrix form as follows:

$$\underset{2\times 2}{\underline{B}}\;\underset{2\times 1}{\underline{\eta}} = \underset{2\times k}{\underline{\Gamma}}\;\underset{k\times 1}{\underline{x}} + \underset{2\times 1}{\underline{\varepsilon}}, \qquad (1.2)$$

where $\underline{x} = (\underline{x}^{(1)T}, \underline{x}^{(2)T}, \underline{x}^{(3)T})^T$. $\underline{x}^{(1)}$ are the variables common to both equations. $\underline{x}^{(2)}$ and $\underline{x}^{(3)}$ are the variables in the first and respectively in the second equation only. $\underline{\Gamma}$ is partitioned accordingly:

$$\underline{\Gamma} = \begin{bmatrix} \underline{\gamma}_{11} & \underline{\gamma}_{12} & \underline{0} \\ \underline{\gamma}_{21} & \underline{0} & \underline{\gamma}_{23} \end{bmatrix}, \quad \begin{matrix} \underline{\gamma}_1^T = (\underline{\gamma}_{11},\underline{\gamma}_{12}) \\ \underline{\gamma}_2^T = (\underline{\gamma}_{21},\underline{\gamma}_{23}) \end{matrix},$$

$$\underline{\eta} = \begin{bmatrix} \eta_1 \\ \eta_2 \end{bmatrix}, \quad \text{and}$$

$$\underline{B} = \begin{bmatrix} 1 & -\beta_1 \\ -\beta_2 & 1 \end{bmatrix} \quad \text{is a regular matrix,}$$

The reduced form of the simultaneous equation system is:

$$\underline{\eta} = \underline{\Pi}\underline{x} + \underline{\zeta}, \qquad (1.3)$$

where $\underline{\Pi} = \underline{B}^{-1}\underline{\Gamma}$, $\underline{\zeta} \sim N_2(\underline{0},\underline{\Omega})$ and $\underline{\Omega} = \underline{B}^{-1}\underline{\Sigma}(\underline{B}^{-1})^T$.

The latent endogenous variable η_1 is measured indirectly by I polytomous indicators Y_i with c_i categories. The indicators of the other latent variable may in principle be of any scale type. At present, however, we assume η_2 to be observed directly. Given a fixed value of the latent variable η_1 the probability of category l of manifest variable Y_i is assumed to be of the multinomial logit type:

$$P(Y_i=1|\eta_1,\underline{\alpha}_i,\underline{\lambda}_i) = P(y_i|\eta_1,\underline{\alpha}_i,\underline{\lambda}_i) = \frac{\exp(\alpha_{il}+\lambda_{il}\eta_1)}{\Sigma \exp(\alpha_{ij}+\lambda_{ij}\eta_1)}, \quad (1.4)$$

where the sum extends from 1 to c_i. Identifying restrictions are: $\alpha_{i1} = \lambda_{i1} = 0$; $i = 1,\ldots,I$.

In principle $P(Y_i=1|\eta_1)$ may be any type of selection probability, especially the multinomial probit model based on the concept of random utility maximization in econometrics (McFadden, 1981) or the equivalent extremal concept in psychometrics (Bock, 1975). The multinomial probit leads to considerable numerical difficulties. But, if independence of errors in the utility functions is assumed, the multinomial probit is well approximated by the multinomial logit (Bock, 1975). Additionally, we assume conditional (local) independence between the indicators Y_i, given a fixed value of η_1 (Bartholomew, 1984). The joint density function is then

$$P(\underline{y}|\eta_1,\underline{A},\underline{\Lambda}) = \prod_{i=1}^{I} P(y_i|\eta_1,\underline{\alpha}_1,\underline{\lambda}_i) \quad (1.5)$$

where $\underline{A}$ is a collection of coefficients $(\underline{\alpha}_1,\ldots \underline{\alpha}_I)$ and $\underline{\Lambda}$ is $(\underline{\lambda}_1,\ldots \underline{\lambda}_I)$. If η_1 and η_2 are observed, the density of the complete data $(\underline{y},\eta_1,\eta_2)$ follows immediately:

$$P(\underline{y},\eta_1,\eta_2|\underline{x},\underline{A},\underline{\Lambda},\underline{B},\underline{\Gamma},\underline{\Sigma}) = P(\underline{y},\eta_1,\eta_2|\underline{x},\underline{A},\underline{\Lambda},\underline{\Pi},\underline{\Omega}) = \quad (1.6)$$

$$P(\underline{y}|\eta_1,\underline{A},\underline{\Lambda})\phi(\eta_1,\eta_2|\underline{x},\underline{\Pi},\underline{\Omega}), \text{ where}$$

$$\phi(\eta_1,\eta_2|\underline{x},\underline{\Pi},\underline{\Omega}) = (2\pi)^{-1}|\underline{\Omega}|^{-\frac{1}{2}}\exp(-\tfrac{1}{2}(\underline{\eta}-\underline{\Pi}\underline{x})^T\underline{\Omega}^{-1}(\underline{\eta}-\underline{\Pi}\underline{x}))$$

The corresponding incomplete data density is obtained by integrating over the latent variable η_1 resulting in the mixture density:

$$P(\underline{y},\eta_2|\underline{x},\underline{A},\underline{\Lambda},\underline{\Pi},\underline{\Omega}) = \int_{\mathbb{R}} P(\underline{y},\eta_1,\eta_2|\underline{x},\underline{A},\underline{\Lambda},\underline{\Pi},\underline{\Omega})d\eta_1 \quad (1.7)$$

As in ordinary probit analysis, a scale restriction must be imposed on

η_1. Hence the variance of η_1 is set to 1. Since the estimation technique proposed herein uses single equation methods, the marginal densities must be obtained. Thus, $\underline{\Pi}$ is partitioned as $\underline{\Pi}_{1.}$ and $\underline{\Pi}_{2.}$ for the first and second equations, respectively. Using the variance restriction for η_1, $\underline{\Omega}$ may be written as:

$$\underline{\Omega} = \begin{bmatrix} 1 & \rho\omega \\ \rho\omega & \omega^2 \end{bmatrix}$$

The complete data marginal density of $(\underline{y},\eta_1)$ is:

$$P(\underline{y},\eta_1|\underline{x},\underline{A},\underline{\Lambda},\underline{\Pi}_{1.}) = P(\underline{y}|\eta_1,\underline{A},\underline{\Lambda})\phi(\eta_1|\underline{x},\underline{\Pi}_{1.}), \tag{1.8}$$

$$\phi(\eta_1|\underline{x},\underline{\Pi}_{1.}) = (2\pi)^{-\frac{1}{2}} \exp(-\tfrac{1}{2}(\eta_1-\underline{\Pi}_{1.}\underline{x})^2).$$

The corresponding incomplete marginal density is therefore:

$$P(\underline{y}|\underline{x},\underline{A},\underline{\Lambda},\underline{\Pi}_{1.}) = \int_{\mathbb{R}} P(\underline{y}|\eta_1,\underline{A},\underline{\Lambda})\phi(\eta_1|\underline{x},\underline{\Pi}_{1.})d\eta_1. \tag{1.9}$$

Since η_2 is directly observed it is marginally normally distributed:

$$\phi(\eta_2|\underline{x},\underline{\Pi}_{2.},\omega^2) = (2\pi\omega^2)^{-\frac{1}{2}} \exp(-\frac{1}{2\omega^2}(\eta_2-\underline{\Pi}_{2.}\underline{x})^2). \tag{1.10}$$

Assuming a sample of independent drawings $\{(\underline{y}_t,\eta_{2t},\underline{x}_t) : t=1,\ldots,T\}$ with stochastic or fixed exogenous variables $\underline{x}_t$ (Anderson and Philips, 1981), the kernel of the incomplete data loglikelihood function is:

$$l(\underline{A},\underline{\Lambda},\underline{B},\underline{\Gamma},\underline{\Sigma}) = l(\underline{A},\underline{\Lambda},\underline{\Pi},\underline{\Omega}) = \sum_{t=1}^{T} \ln P(\underline{y}_t,\eta_{2t}|\underline{x}_t,\underline{A},\underline{\Lambda},\underline{\Pi},\underline{\Omega}) \tag{1.11}$$

The parameters $\underline{A}$, $\underline{\Lambda}$, $\underline{B}$, $\underline{\Gamma}$, and $\underline{\Sigma}$ may be estimated directly from equation (1.11) by full information maximum likelihood. To avoid numerical complications arising in large systems, a systematic marginal and sequential likelihood estimation strategy is used.

2. ESTIMATION OF THE LATENT TRAIT MODEL

2.1 Ordinary ML-Estimation

The basis for statistical inference for the parameters $\underline{A}$, $\underline{\Lambda}$, $\underline{\Pi}_{1.}$ of the first equation is the incomplete data marginal loglikelihood:

$$l(\underline{A},\underline{\Lambda},\underline{\Pi}_{1.}) = \sum_{t=1}^{T} \ln P(\underline{y}_t|\underline{x}_t,\underline{A},\underline{\Lambda},\underline{\Pi}_{1.}) = \sum_{t=1}^{T} \ln P_t \; . \tag{2.1}$$

A solution of the likelihood equation may be obtained by iterative numerical methods using the first derivatives of $l(\underline{A},\underline{\Lambda},\Pi_{1.})$ only.

The following notation simplifies the derivations:

$$\underline{\delta}_i = (\underline{\delta}_{i1}^T,\ldots,\underline{\delta}_{ic_i}^T)^T,\; \underline{\delta}_{il}^T = (\alpha_{il},\lambda_{il}),\; \tilde{\underline{\eta}}_1 = (1,\eta_1)^T,$$

$$\underline{\theta} = (\underline{\delta}_1^T,\ldots,\underline{\delta}_I^T,\underline{\Pi}_{1.})^T,\; \phi_t = \phi(\eta_1|\underline{x}_t,\underline{\Pi}_{1.}),$$

$$P_{ti} = P(y_{ti}|\eta_1,\underline{\alpha}_i,\underline{\lambda}_i),\; P_{ti}(l) = P(y_{ti}=l|\eta_1,\underline{\alpha}_i,\underline{\lambda}_i),$$

$$i = 1,\ldots,I;\; l = 1,\ldots,c_i;\; t = 1,\ldots,T.$$

The first derivatives are given by

$$\frac{\partial l}{\partial \underline{\theta}} = \sum_{t=1}^{T} \frac{1}{P_t} \frac{\partial P_t}{\partial \underline{\theta}} \quad \text{where} \tag{2.2}$$

$$\frac{\partial P_t}{\partial \underline{\theta}} = \int_{-\infty}^{\infty} \frac{\partial}{\partial \underline{\theta}}\left\{\left[\prod_{i=1}^{I} P(y_{ti}|\eta_1,\underline{\alpha}_i,\underline{\lambda}_i)\right]\phi(\eta_i|\underline{x}_t,\underline{\Pi}_{1.})\right\}d\eta_1 .$$

The derivatives with respect to $\underline{\delta}_j$ under the integral are collected in the following vectors:

$$\partial(\phi_t \prod_{i=1}^{I} P_{ti})/\partial\underline{\delta}_j = \phi_t(\partial P_{tj}/\partial\underline{\delta}_j)(\prod_{i\neq j}^{I} P_{ti}). \tag{2.3}$$

The elements of the vectors are:

$$\partial P_{tj}(l)/\partial \underline{\delta}_{jk} = -P_{tj}(l)P_{tj}(k)\underline{\tilde{\eta}}_1, \ l=1,\ldots,c_j, \ k=2,\ldots,c_j, \ k\neq l. \quad (2.4)$$

$$\partial P_{tj}(l)/\partial \underline{\delta}_{jl} = (P_{tj}(l)-P_{tj}(l)^2)\underline{\tilde{\eta}}_1, \ l=2,\ldots,c_j.$$

$$\partial P_{tj}(l)/\partial \underline{\delta}_{j'k} = 0, \quad \text{if} \quad j\neq j'.$$

The derivative with respect to $\underline{\Pi}_{1.}$ is

$$\partial(\phi_t \prod_{i=1}^{I} P_{ti})/\partial \underline{\Pi}_{1.}^T = (\partial\phi_t/\partial\underline{\Pi}_{1.}^T)(\prod_{i=1}^{I} P_{ti}), \quad (2.5)$$

where $\partial\phi_t/\partial\underline{\Pi}_{1.}^T = (\eta_1-\underline{\Pi}_{1.}\underline{x}_t)\underline{x}_t\phi_t$.

Computing the first derivatives from equation (2.2) involves numerical integration of each term P_t and $\partial P_t/\partial\underline{\theta}$. Both P_t and $\partial P_t/\partial\underline{\theta}$ involve ϕ_t and are therefore of the form:

$$\int_{-\infty}^{\infty} f(\eta_1)\exp(-\tfrac{1}{2}(\eta_1-\underline{\Pi}_{1.}\underline{x}_t)^2)d\eta_1 =$$

$$\int_{-\infty}^{\infty} f(\sqrt{2}\,\xi+\underline{\Pi}_{1.}\underline{x}_t)\exp(-\xi^2)\sqrt{2}\,d\xi. \quad (2.6)$$

Using Gauß-Hermite integration with support points ξ_h and weights w_h, $h=1,\ldots,H$ (Stroud and Secrest, 1966) yields:

$$\sum_{h=1}^{H} f(\sqrt{2}\,\xi_h+\underline{\Pi}_{1.}\underline{x}_t)\sqrt{2}\,w_h. \quad (2.7)$$

Estimates of the asymptotic covariance matrix $\underline{V}(\hat{\underline{\theta}})$ are usually computed using the inverse of the Fisher information matrix or the inverse of the observed information matrix. Since the second derivatives are cumbersome to compute, it is convenient to approximate the expected information matrix $\underline{I}(\underline{\theta})$ by its empirical counterpart, evaluated at the solution of the likelihood equations $\hat{\underline{\theta}}$:

$$\underline{I}(\hat{\underline{\theta}}) = \frac{1}{T} \sum_{t=1}^{T} \frac{1}{P_t^2} \left(\frac{\partial P_t}{\partial \underline{\theta}}\right)\left(\frac{\partial P_t}{\partial \underline{\theta}}\right)^T \tag{2.8}$$

2.2 ML-Estimation with the EM-algorithm

In the EM-algorithm (Dempster et al., 1977) the expected value of the logarithm of the complete data likelihood conditional on the observed data and the parameter estimates from the previous iteration is maximized:

$$Q(\underline{\theta}^{q+1}|\underline{\theta}^q) = \sum_{t=1}^{T} E\{\ln P(\underline{y}_t, \eta_{1t}|\underline{x}_t, \underline{A}^{q+1}, \underline{\Lambda}^{q+1}, \underline{\Pi}_{1.}^{q+1})|\underline{y}_t, \underline{\theta}^q\} \tag{2.9}$$

The conditional density $(\underline{y}_t, \eta_1|\underline{x}_t)$ given $\underline{y}_t$ and $\underline{\theta}^q$ is:

$$g(\underline{y}_t, \eta_1|\underline{x}_t, \underline{y}_t, \underline{\theta}^q) = \frac{P(\underline{y}_t|\eta_1, \underline{A}^q, \underline{\Lambda}^q)\phi(\eta_1|\underline{x}_t, \underline{\Pi}_{1.}^q)}{\int_{-\infty}^{\infty} P(\underline{y}_t|\eta_1, \underline{A}^q, \underline{\Lambda}^q)\phi(\eta_1|\underline{x}_t, \underline{\Pi}_{1.}^q)d\eta_1} \quad . \tag{2.10}$$

Hence $Q(\underline{\theta}^{q+1}|\underline{\theta}^q)$ may be written as

$$\sum_{t=1}^{T} \int_{-\infty}^{\infty} \ln\left[P(\underline{y}_t|\eta_1, \underline{A}^{q+1}, \underline{\Lambda}^{q+1})\phi(\eta_1|\underline{x}_t, \underline{\Pi}_{1.}^{q+1})\right] g(\underline{y}_t, \eta_1|\underline{x}_t, \underline{y}_t, \underline{\theta}^q)d\eta_1 = \tag{2.11}$$

$$\sum_{i=1}^{I} \sum_{t=1}^{T} \int_{-\infty}^{\infty} \ln\left[P(y_{ti}|\eta_1, \underline{\alpha}_i^{q+1}, \underline{\lambda}_i^{q+1})\right] g(\underline{y}_t, \eta_1|\underline{x}_t, \underline{y}_t, \underline{\theta}^q)d\eta_1 +$$

$$\sum_{t=1}^{T} \int_{-\infty}^{\infty} \ln\left[\phi(\eta_1|\underline{x}_t, \underline{\Pi}_{1.}^{q+1})\right] g(\underline{y}_t, \eta_1|\underline{x}_t, \underline{y}_t, \underline{\theta}^q)d\eta_1 =$$

$$\sum_{i=1}^{I} Q_i(\underline{\alpha}_i^{q+1}, \underline{\lambda}_i^{q+1}|\underline{\theta}^q) \quad + \quad Q_o(\underline{\Pi}_{i.}^{q+1}|\underline{\theta}^q) \; .$$

As may be seen from the last line, Q may be maximized by optimizing I + 1 separate terms, each involving the posterior density $g(\underline{y}_t, \eta_1|\underline{x}_t, \underline{y}_t, \underline{\theta}^q)$. The denominator of this density must be computed by numerical integration using (2.6). The denominator of equation (2.10) is denoted by d_t. Again using numerical integration, each Q_i $i=1,...,I$ may be written as:

$$Q_i = \sum_{t=1}^{T} \frac{1}{d_t} \int_{-\infty}^{\infty} \ln\left[P(y_{ti}|\eta_1,\underline{\alpha}_i^{q+1},\underline{\lambda}_i^{q+1})\right] P(\underline{y}_t|\eta_1,\underline{A}^q,\underline{\Lambda}^q)\phi(\eta_1|\underline{x}_t,\underline{\Pi}_{1.}^q)d\eta_1 \cong$$

$$\sum_{t=1}^{T} \frac{1}{d_t} \sum_{h=1}^{H} \ln\left[P(y_{ti}|\sqrt{2}\,\xi_h+\underline{\pi}_{1.}^q\,\underline{x}_t,\underline{\alpha}_i^{q+1},\underline{\lambda}_i^{q+1})\right] *$$

$$P(\underline{y}_t|\sqrt{2}\,\xi_h+\underline{\Pi}_{1.}^q\,\underline{x}_t,\underline{A}^q,\underline{\Lambda}^q)\;\frac{w_h}{\sqrt{\pi}} \qquad (2.12)$$

The maximization of Q_i corresponds to a sequence of multinomial logit maximizations with TxH cases and known weights v_{th}, where

$$v_{th} = (w_h/(d_t\sqrt{\pi}))\; P(\underline{y}_t|\sqrt{2}\,\xi_h+\underline{\Pi}_{1.}^q\,\underline{x}_t,\underline{A}^q,\underline{\Lambda}^q). \qquad (2.13)$$

The logit regressor η_1 is replaced by $\sqrt{2}\,\xi_h + \underline{\Pi}_{1.}^q\,\underline{x}_t$. The same procedure is applied to compute Q_o:

$$Q_o = \sum_{t=1}^{T}\sum_{h=1}^{H} \ln\,\phi(\sqrt{2}\,\xi_h+\underline{\Pi}_{1.}^q\,\underline{x}_t|\underline{x}_t,\underline{\Pi}_{1.}^{q+1})v_{th}. \qquad (2.14)$$

Maximizing Q_o is equivalent to a sequence of weighted regressions with TxH cases and known weights v_{th}. An estimator of the observed information matrix may in principle be obtained from the formulae of Louis (1982). However, this procedure is computationally burdensome.

2.3 Computing initial estimates with GLIM

Initial estimates of the parameters may be computed with GLIM in three stages in each M-step of the EM-algorithm: First, for each case and each support point the values of the weights v_{th} are calculated from equation (2.13). Second, the vector of the reduced form coefficients $\underline{\Pi}_{1.}^{q+1}$ is estimated by least squares with known weights v_{th}. Third, for each indicator i=1,...,I, the parameters $\underline{\alpha}_i$ and $\underline{\lambda}_i$ are approximated by the following procedure modifying the results of Begg and Gray (1984). The multinomial logit model in Q_i is broken up into c_i-1 binomial logits for each case and each support point using the relationship:

$$P(y_{ti}=l|y_{ti}\in\{1,l\},\ \alpha_{il},\lambda_{il},\sqrt{2}\,\xi_h+\underline{\Pi}_{1.}\,\underline{x}_t) \qquad (2.15)$$

$$= \exp(\alpha_{il}+\lambda_{il}\eta_{th})/(1+\exp(\alpha_{il}+\lambda_{il}\eta_{th})),$$

where $\eta_{th} = \sqrt{2}\,\xi_h + \underline{\Pi}^q_{1.}\underline{x}_t$.

This is equivalent to considering binary odds with choice sets {1,l}, l=2,... c_i. Hence, the following term is maximized for each combination (1,l), l=2,... c_i:

$$\sum_{t=1}^{T} \sum_{h=1}^{H} u_t v_{th} \ln P(y_{ti} | y_{ti} \epsilon \{1,l\}, \alpha_{il}, \lambda_{il}, \eta_{th}), \qquad (2.16)$$

where $u_t = 1$, if $y_{ti} \in \{1,l\}$, and $u_t = 0$, otherwise. This maximization is easily performed by using the binomial logit model and the weight directive in GLIM (Baker and Nelder, 1978).

2.4 Latent trait score estimation

The η_1 score of case t can be estimated in two ways. In either case we consider the posterior density $g(\underline{y}_t, \eta_1 | \underline{x}_t, \hat{\underline{\theta}})$ of equation (2.10) evaluated at the final estimate $\hat{\underline{\theta}}$. A first estimate is obtained as when computing factor scores, by considering η_1 as a case specific parameter η_{1t} and maximizing $g(\underline{y}_t, \eta_{1t} | \underline{x}_t, \underline{y}_t, \hat{\underline{\theta}})$ with respect to η_{1t}. Maximization is straightforward using first derivatives and corresponding iterative optimization routines.

A second estimate is based on the fact that η_1 is a random variable (Cf. Bartholomew, 1981) with density given in equation (2.10) conditional on $\underline{x}_t, \underline{y}_t$ and the true parameters $\underline{\theta}$. If $\underline{\theta}$ is replaced by $\hat{\underline{\theta}}$, it is possible to estimate the latent score as the expected value of $(\eta_1 | \underline{x}_t, \underline{y}_t, \hat{\underline{\theta}})$. This results in the evaluation of the integral:

$$\hat{\eta}_{1t} = E(\eta_1 | \underline{x}_t, \underline{y}_t, \hat{\underline{\theta}}) = \int_{-\infty}^{\infty} \eta_1 \; g(\underline{y}_t, \eta_1 | \underline{x}_t, \underline{y}_t, \hat{\underline{\theta}}) d\eta_1. \qquad (2.17)$$

Using numerical integration, the integral is approximated by:

$$\hat{\eta}_{1t} \cong \sum_{h=1}^{H} (\sqrt{2}\,\xi_h + \underline{\Pi}_{1.}\underline{x}_t) \hat{v}_{th},$$

where $\hat{v}_{th} = v_{th}$ evaluated at $\hat{\underline{\theta}}$.

3. ESTIMATION OF THE REDUCED FORM CORRELATION

Marginal likelihood estimation of $\underline{\Pi}_{2.}$ and ω^2 in the second equation of the simultaneous equation system is easily achieved by maximization of

$$\prod_{t=1}^{T} \phi(\eta_{2t} | \underline{x}_t, \underline{\Pi}_{2.}, \omega^2), \tag{3.1}$$

yielding estimators $\hat{\underline{\Pi}}_{2.}$ and $\hat{\omega}^2$.

Our next goal is the estimation of ρ in $\underline{\Omega}$. For this purpose we use the latent trait score residuals $\hat{\zeta}_{1t} = \hat{\eta}_{1t} - \hat{\underline{\Pi}}_{1.}\underline{x}_t$, the ordinary least squares residuals $\hat{\zeta}_{2t} = \eta_{2t} - \hat{\underline{\Pi}}_{2.}\underline{x}_t$ and the estimated variance $\hat{\omega}^2$, yielding

$$\hat{\rho} = (T\hat{\omega})^{-1} \sum_{t=1}^{T} \hat{\zeta}_{1t}\hat{\zeta}_{2t}. \tag{3.2}$$

4. ESTIMATION OF THE STRUCTURAL PARAMETERS

To derive estimators of $\underline{B}$ and $\underline{\Gamma}$ from the reduced form estimators $\hat{\underline{\Pi}}$ we utilize Amemiya's (1979) principle. This principle consists in writing the identity $\underline{\Gamma} = \underline{B}\underline{\Pi}$ in its equations components

$$(\gamma_{11}, \gamma_{12}, 0) = \Pi_{1.} - \beta_1 \Pi_{2.}, \tag{4.1}$$

$$(\gamma_{21}, 0, \gamma_{23}) = \Pi_{2.} - \beta_2 \Pi_{1.}.$$

Inserting $\Pi = \underline{\Pi} + \hat{\underline{\Pi}} - \hat{\underline{\Pi}}$ into equations (4.1) results in:

$$\hat{\underline{\Pi}}_{1.} = \beta_1 \hat{\underline{\Pi}}_{2.} + (\underline{\gamma}_{11}, \underline{\gamma}_{12}, \underline{0}) + \underline{\delta}_1, \tag{4.2}$$

$$\hat{\underline{\Pi}}_{2.} = \beta_2 \hat{\underline{\Pi}}_{1} + (\underline{\gamma}_{21}, 0, \underline{\gamma}_{23}) + \underline{\delta}_2,$$

where $\underline{\delta}_1 = (\hat{\underline{\Pi}}_{1.} - \underline{\Pi}_{1.}) + \beta_1(\underline{\Pi}_{2.} - \hat{\underline{\Pi}}_{2.})$,

$\underline{\delta}_2 = (\hat{\underline{\Pi}}_{2.} - \underline{\Pi}_{2.}) + \beta_2(\underline{\Pi}_{1.} - \hat{\underline{\Pi}}_{1.})$.

The parameter vectors $(\beta_1, \underline{\gamma}_{11}, \underline{\gamma}_{12})$ and $(\beta_2, \underline{\gamma}_{21}, \underline{\gamma}_{23})$ are denoted by $\underline{b}_1$ and $\underline{b}_2$. Let $\underline{I}_1$, $\underline{I}_2$, $\underline{I}_3$ be identity matrices with orders corresponding to $\underline{\gamma}_{11}$, $\underline{\gamma}_{12}$, $\underline{\gamma}_{23}$. Then equation (4.2) may be rewritten as:

$$\underline{\hat{\Pi}}_i = \underline{b}_i \underline{Z}_i + \underline{\delta}_i, \quad i=1,2, \tag{4.3}$$

where $\underline{Z}_1 = \begin{bmatrix} & \underline{\hat{\Pi}}_{2.} & \\ \underline{I}_1 & \underline{0} & \underline{0} \\ \underline{0} & \underline{I}_2 & \underline{0} \end{bmatrix}$, $\underline{Z}_2 = \begin{bmatrix} & \underline{\hat{\Pi}}_{1.} & \\ \underline{I}_1 & \underline{0} & \underline{0} \\ \underline{0} & \underline{0} & \underline{I}_3 \end{bmatrix}$.

Since $\underline{\hat{\Pi}}_i$ and $\underline{Z}_i$ are obtained in the previous estimation stages, $\underline{b}_1$ can be estimated immediately from equation (4.3) by ordinary least squares:

$$\underline{\hat{b}}_i = \underline{\hat{\Pi}}_{i.} \underline{Z}_i^T (\underline{Z}_i \underline{Z}_i^T)^{-1} . \tag{4.4}$$

Note that the calculations of $\underline{\hat{b}}_i$ is - compared to similar two stage estimators (Cf Maddala, 1983) - easily performed because of the low column order of $\underline{Z}_i$. Having obtained an estimator of $\underline{B}$ and $\underline{\Omega}$ the covariance $\underline{\Sigma}$ of the simultaneous equation system is computed immediately:

$$\underline{\hat{\Sigma}} = \underline{\hat{B}}\underline{\hat{\Omega}}\underline{\hat{B}}^T. \tag{4.5}$$

5. SOME EXTENSIONS

The model above may be extended in a theoretically straightforward but computationally tedious manner in four ways. First, it is possible to consider three or more equations simultaneously. As before, Amemiya's principle may be used to estimate the structural parameters, though the formulae for this case are slightly more complicated. Second, each latent variable may be measured by either categorical, ordinal, censored or metric indicators. Third, the indicators for each latent variable may be of a mixed measurement level. Fourth, there may be more than one factor for each set of indicators. In this case it is necessary to evaluate multiple integrals numerically.

REFERENCES

Amemiya, T. (1979). The Estimation of a Simultaneous-Equation Tobit Model. International Economic Review 20, 169 - 181.

Anderson, J.A. and Philips, P.R. (1981). Regression, Discrimination and Measurement Model for Ordered Categorical Variables. Applied Statistics 30, 22 - 31.

Baker, R.J. and Nelder, J.A. (1978). The GLIM system, Release 3. Numerical Algorithm Group, Oxford.

Bartholomew, D.J. (1981) Posterior analysis of the factor model. British Journal of Mathematical and Statistical Psychology 34, 93 - 99.

Bartholomew, D.J. (1984) The foundations of factor analysis. Biometrika 71, 221 - 232.

Begg, C.B. and Gray, R. (1984) Calculation of polychotomous logistic regression parameters using individualized regressions. Biometrika 71, 11 - 18.

Bock, R.D. (1972) Estimating Item Parameters and Latent Ability when Responses are Scored in two or more Nominal Categories. Psychometrika 37, 29 - 51.

Bock, R.D. (1975), Multivariate Statistical Methods in Behavioral Research, New York: McGraw-Hill.

Dempster, A.P., N.M. Laird and Rubin, D.B. (1977) Maximum Likelihood from Incomplete Data via the EM-Algorithm. Journal of the Royal Statistical Society Series B, 39, 1 - 38.

Jøreskog, K.G. and Sørbom, D. (1984) LISREL VI - Analysis of Linear Structural Relationship by Maximum-Likelihood, Instrumental Variables and Least Squares Method. Scientific Software Inc., Mooresville, Indiana.

Louis, T.A. (1982) Finding the Observed Information Matrix when using the EM-Algorithm. Journal of the Royal Statistical Society, Series B 44, 226 - 233.

Maddala, G.S. (1983) Limited-dependent and qualitative variable in econometrics. Cambridge: Cambridge University Press.

McFadden, D. (1981) Econometric Models of Probabilitstic Choice. In: Manski, Ch.F. and McFadden, D. (eds.) (1981). Structural Analysis of Discrete Data with Econometric Applications, 198 - 272. Cambridge: MIT-Press.

Muthên, B. (1984). A General Structural Equation Model with Dichotomous, Ordered Categorical and Continous Latent Variable Indicators. Psychometrika 49, 115 - 132.

Schmidt, P. (1976). Econometrics. New York - Basel: Marcel Dekker.

Stroud, A. and Secrest, D. (1966) Gaussian Quadrature Formulas. Englewood Cliffs - New York: Prentice-Hall.

THE NULL EXPECTED DEVIANCE FOR AN EXTENDED CLASS OF GENERALIZED LINEAR MODELS

By Gauss M Cordeiro
CCEN/UFPE, Cidade Universitária
Recife 50.000 - PE - Brazil

ABSTRACT

Jørgensen (1983) developed a class of extended generalized linear models including error distributions not of the exponential family form. We give the null expected likelihood ratio statistic up to order n^{-1}, where n is the sample size, for testing the goodness-of-fit of an extended generalized linear model. This result generalizes previous work by Cordeiro (1983). Applications to several models are discussed.

Keywords: expected deviance, extended generalized linear model, likelihood ratio, link function.

1. INTRODUCTION

Jørgensen (1983) defines an extended class of generalized linear models by assuming that $Y_1 \ldots Y_n$ are independent random variables and that the probability density function of each Y_ℓ has the form

$$\Pi(y;\, \theta_\ell,\, \phi_\ell) = \exp\{p(\phi_\ell)\; t(y,\theta_\ell) + q(y,\phi_\ell)\} \tag{1.1}$$

where $p(\cdot)$, $t(\cdot,\cdot)$ and $q(\cdot,\cdot)$ are known fuctions and $p(\phi_\ell) > 0$, $\ell=1\ldots n$.

The systematic part of the model is specified by a linear structure $\underset{\sim}{\eta} = \underset{\sim}{X}\,\underset{\sim}{\beta}$, where $\underset{\sim}{\eta} = (\eta_1 \ldots \eta_n)^T$, $X = \{x_{\ell i}\}$ is an nxp known model matrix of full rank $p \leqslant n$ and $\underset{\sim}{\beta} = (\beta_1 \ldots \beta_p)^T$ is a set of p unknown parameters, and by a function $\theta_\ell = F(\eta_\ell)$, $\ell=1\ldots n$, connecting (1.1) and the linear structure. The function $F(\cdot)$ is assumed to be one-to-one and differentiable. Jørgensen showed that these models have several properties similar to those of the generalized linear models. The method of maximum likelihood is used to estimate the linear parameters $\beta_1 \ldots \beta_p$ and hence $\theta_1 \ldots \theta_n$. Throughout the paper, the parameter ϕ is assumed known for each observation and therefore any density with only one unknown parameter may be written in the form (1.1). Let $y_1 \ldots y_n$ be the observations and suppose that the log likelihood function for the extended generalized linear model is $L(\underset{\sim}{\beta}) = \sum_{\ell=1}^{n} \log \Pi(y_\ell;\theta_\ell,\phi_\ell)$. Let $\hat{\underset{\sim}{\beta}}$, $\hat{\underset{\sim}{\eta}}$ and $\hat{\underset{\sim}{\theta}}$ be the maximum likelihood estimates of $\underset{\sim}{\beta}$, $\underset{\sim}{\eta}$ and $\underset{\sim}{\theta}$ respectively. We define the following expectations $D_i(\theta) = E\left(\frac{\partial^i t(Y,\theta)}{\partial\theta^i}\right)$, $i = 1, 2, \ldots$.

On differentiating $L(\underset{\sim}{\beta})$ and taking expectation we obtain $k_{rs} = E\left(\frac{\partial^2 L(\underset{\sim}{\beta})}{\partial\beta_r\partial\beta_s}\right) = -\sum_{\ell=1}^{n} p(\phi_\ell)\, w_\ell\, x_{\ell r}\, x_{\ell s}$ where $w = -D_2(\theta)\left(\frac{d\theta}{d\eta}\right)$. The

Fisher information matrix is pxp and is given by $\underset{\sim}{K} = \{-k_{rs}\} = \underset{\sim}{X}^T \underset{\sim}{W} \underset{\sim}{\phi} \underset{\sim}{X}$ where $\underset{\sim}{W} = \text{diag}\{w_1 \ldots w_n\}$ and $\underset{\sim}{\phi} = \text{diag}\{p(\phi_1) \ldots p(\phi_n)\}$ are diagonal matrices of order n. Because X is assumed full rank and w and $p(\phi)$ are positive, the information matrix is positive definite and so is its inverse $\underset{\sim}{K}^{-1} = \{-k^{rs}\}$.

2. IMPROVED LIKELIHOOD RATIO TESTS FOR EXTENDED GENERALIZED LINEAR MODELS

In this section we give the n^{-1} term in the null expected likelihood ratio statistic for the extended generalized linear model. This result generalizes previous work by Cordeiro (1983).

We use the notation $k_{rst} = E\left(\frac{\partial^3 L(\underset{\sim}{\beta})}{\partial\beta_r \partial\beta_s \partial\beta_t}\right)$, $k_{rs}^{(t)} = \frac{\partial k_{rs}}{\partial \beta_t}$,

$k_{rs}^{(tu)} = \frac{\partial^2 k_{rs}}{\partial\beta_t \partial\beta_u}$ and so on. The k's refer to a total over the sample and are, in general, of order n. Denote by L the value of $L(\underset{\sim}{\beta})$ at the true parameter point. Lawley (1956) showed under general regularity conditions for large - sample theory that $2\,E\left[L(\hat{\underset{\sim}{\beta}}) - L\right] = p + \varepsilon_p + O(n^{-2})$ where

$$\varepsilon_p = \sum_{r,s,t,u} k^{rs}k^{tu}\ell_{rstu} - \sum_{r,s,t,u,v,w} k^{rs}k^{tu}k^{vw}\ell_{rstuvw} \tag{2.1}$$

is of order n^{-1} with

$\ell_{rstu} = \frac{1}{4}k_{rstu} - k_{rst}^{(u)} + k_{rt}^{(su)}$, $\ell_{rstuvw} = k_{rtv}\left(\frac{1}{6}k_{suw} - k_{sw}^{(u)}\right) + k_{rt}^{(v)}k_{sw}^{(u)} + k_{rtu}\left(\frac{1}{4}k_{svw} - k_{sw}^{(v)}\right) + k_{rt}^{(u)}k_{sw}^{(v)}$ and all k's being evaluated at the true point. Differentiating k_{rs} yields

$$k_{rs}^{(t)} = \sum_{\ell=1}^{n} p(\phi_\ell)\left[\left(\frac{d\theta}{d\eta}\right)^3 \frac{dD_2(\theta)}{d\theta} + 2\frac{d\theta}{d\eta}\frac{d^2\theta}{d\eta^2}D_2(\theta)\right]_\ell x_{\ell r}x_{\ell s}x_{\ell t} \tag{2.2}$$

and

$$k_{rt}^{(su)} = \sum_{\ell=1}^{n} p(\phi_\ell)\left[\left(\frac{d\theta}{d\eta}\right)^4 \frac{d^2D_2(\theta)}{d\theta^2} + 5\left(\frac{d\theta}{d\eta}\right)^2\frac{d^2\theta}{d\eta^2}\frac{dD_2(\theta)}{d\theta} + 2\left\{\left(\frac{d^2\theta}{d\eta^2}\right)^2 + \frac{d\theta}{d\eta}\frac{d^3\theta}{d\eta^3}\right\}D_2(\theta)\right]_\ell x_{\ell r}x_{\ell s}x_{\ell t}x_{\ell u}. \tag{2.3}$$

For the expected third and fourth derivatives of $L(\underset{\sim}{\beta})$ we find, respectively,

$$k_{rst} = \sum_{\ell=1}^{n} p(\phi_\ell)\left[3\frac{d\theta}{d\eta}\frac{d^2\theta}{d\eta^2}D_2(\theta) + \left(\frac{d\theta}{d\eta}\right)^3 D_3(\theta)\right]_\ell x_{\ell r}x_{\ell s}x_{\ell t} \tag{2.4}$$

and

$$k_{rstu} = \sum_{\ell=1}^{n} p(\phi_\ell)\left[\left\{4\frac{d\theta}{d\eta}\frac{d^3\theta}{d\eta^3} + 3\left(\frac{d^2\theta}{d\eta^2}\right)^2\right\}D_2(\theta) + 6\left(\frac{d\theta}{d\eta}\right)^2\frac{d^2\theta}{d\eta^2}D_3(\theta) + \left(\frac{d\theta}{d\eta}\right)^4 D_4(\theta)\right]_\ell x_{\ell r}x_{\ell s}x_{\ell t}x_{\ell u}. \tag{2.5}$$

Differentiating (2.4) gives

$$k_{rst}^{(u)} = \sum_{\ell=1}^{n} p(\phi_\ell) \Big[3\Big(\frac{d\theta}{d\eta}\Big)^2 \frac{d^2\theta}{d\eta^2} \frac{dD_2(\theta)}{d\theta} + \Big(\frac{d\theta}{d\eta}\Big)^4 \frac{dD_3(\theta)}{d\theta} +$$

$$3\Big\{\Big(\frac{d^2\theta}{d\eta^2}\Big)^2 + \frac{d^3\theta}{d\eta^3}\frac{d\theta}{d\eta}\Big\} D_2(\theta) + 3\Big(\frac{d\theta}{d\eta}\Big)^2 \frac{d^2\theta}{d\eta^2} D_3(\theta)\Big]_\ell \, x_{\ell r} x_{\ell s} x_{\ell t} x_{\ell u}. \quad (2.6)$$

From (2.3), (2.5) and (2.6) $\ell_{rstu} = \frac{1}{4} \sum_{\ell=1}^{n} p(\phi_\ell)\, h_\ell\, x_{\ell r} x_{\ell s} x_{\ell t} x_{\ell u}$

where $h = 4\Big(\frac{d\theta}{d\eta}\Big)^4 \frac{d^2 D_2(\theta)}{d\theta^2} + 8\Big(\frac{d\theta}{d\eta}\Big)^2 \frac{d^2\theta}{d\eta^2} \frac{dD_2(\theta)}{d\theta} - 4\Big(\frac{d\theta}{d\eta}\Big)^4 \frac{dD_3(\theta)}{d\theta}$

$- \Big(\frac{d^2\theta}{d\eta^2}\Big)^2 D_2(\theta) - 6\Big(\frac{d\theta}{d\eta}\Big)^2 \frac{d^2\theta}{d\eta^2} D_3(\theta) + \Big(\frac{d\theta}{d\eta}\Big)^4 D_4(\theta)$. We define

$g = -\frac{d\theta}{d\eta}\frac{d^2\theta}{d\eta^2} D_2(\theta)$, $e = \Big(\frac{d\theta}{d\eta}\Big)^3 \frac{dD_2(\theta)}{d\theta}$, $f = -\frac{d\theta}{d\eta}\frac{d^2\theta}{d\eta^2} D_2(\theta) - \Big(\frac{d\theta}{d\eta}\Big)^3 D_3(\theta)$

and from (2.2) and (2.4) we get, respectively, $k_{rs}^{(t)} = -\sum_{\ell=1}^{n} p(\phi_\ell)(e_\ell + 2g_\ell)\, x_{\ell r} x_{\ell s} x_{\ell t}$ and $k_{rst} = -\sum_{\ell=1}^{n} p(\phi_\ell)(f_\ell + 2g_\ell)\, x_{\ell r} x_{\ell s} x_{\ell t}$.

Therefore $k_{rtv}\big(\frac{1}{6} k_{suw} - k_{sw}^{(u)}\big) + k_{rt}^{(v)} k_{sw}^{(u)} = \sum_{\ell,m=1}^{n} p(\phi_\ell)\, p(\phi_m)$
$\big[(f_\ell + 2g_\ell)(-e_m - \frac{5}{3} g_m + \frac{1}{6} f_m) + (e_\ell + 2g_\ell)(e_m + 2g_m)\big]\, x_{\ell r} x_{\ell t} x_{\ell v} x_{ms} x_{mu} x_{mw}$ and $k_{rtu}\big(\frac{1}{4} k_{svw} - k_{sw}^{(v)}\big) + k_{rt}^{(u)} k_{sw}^{(v)} =$
$\sum_{\ell,m=1}^{n} p(\phi_\ell) p(\phi_m) \big[f_\ell(\frac{1}{4} f_m - g_m - e_m) + g_m(g_\ell + 2e_\ell) + e_\ell e_m\big] x_{\ell r} x_{\ell t} x_{\ell u} x_{ms} x_{mv} x_{mw}$.

Let $\tilde{Z} = \{z_{\ell m}\} = \tilde{X}(\tilde{X}^T \tilde{W}\tilde{\phi}\tilde{X})^{-1}\tilde{X}^T$, an nxn positive semi-definite matrix of rank p, be the asymptotic covariance matrix for $\hat{\eta}_1 \ldots \hat{\eta}_n$.

Expressions $\sum_{r,s,t,u} k^{rs} k^{tu} \ell_{rstu}$ and $\sum_{r,s,t,u,v,w} k^{rs} k^{tu} k^{vw} \ell_{rstuvw}$ may be found carrying out the sums over the sample after evaluating the sums over the parameters and noting that $-\sum_{r,s} k^{rs} x_{\ell r} x_{ms} = z_{\ell m}$.

Finally, we get the n^{-1} term ε_p by using matrix notation

$$\varepsilon_p = \frac{1}{4}\,\mathrm{tr}(\tilde{\phi}\,\tilde{H}\,\tilde{Z}_d^2) + \tilde{1}^T \tilde{\phi}\,(\tilde{F} + 2\tilde{G})\,\tilde{Z}^{(3)}\Big(\frac{1}{6}\tilde{F} - \frac{5}{3}\tilde{G} - \tilde{E}\Big)\tilde{\phi}\,\tilde{1}$$
$$+ \tilde{1}^T \tilde{\phi}\,(\tilde{E} + 2\tilde{G})\,\tilde{Z}^{(3)}(\tilde{E} + 2\tilde{G})\,\tilde{\phi}\,\tilde{1} - \frac{1}{4}\tilde{1}^T \tilde{\phi}\,\tilde{F}\,\tilde{Z}_d\,\tilde{Z}\,\tilde{Z}_d(4\tilde{G} + 4\tilde{E} - \tilde{F})\,\tilde{\phi}\,\tilde{1}$$
$$+ \tilde{1}^T \tilde{\phi}\,\tilde{G}\,\tilde{Z}_d\,\tilde{Z}\,\tilde{Z}_d(\tilde{G} + 2\tilde{E})\,\tilde{\phi}\,\tilde{1} + \tilde{1}^T \tilde{\phi}\,\tilde{E}\,\tilde{Z}_d\,\tilde{Z}\,\tilde{Z}_d\,\tilde{E}\,\tilde{\phi}\,\tilde{1} \quad (2.7)$$

where $\tilde{E} = \mathrm{diag}\{e_1 \ldots e_n\}$, $\tilde{F} = \mathrm{diag}\{f_1 \ldots f_n\}$, $\tilde{G} = \mathrm{diag}\{g_1 \ldots g_n\}$, $\tilde{H} = \mathrm{diag}\{h_1 \ldots h_n\}$ and $\tilde{Z}_d = \mathrm{diag}\{z_{11} \ldots z_{nn}\}$ are diagonal matrices of order n and $Z^{(3)} = \{z_{\ell m}^3\}$.

The ε_p term depends on the model matrix $\tilde{X}$, on the first four expected derivatives of the function $t(Y,\theta)$ with respect to θ, on the first two derivatives of the function $F(\cdot)$ and on the unknown $\tilde{\beta}$. The dependence on $t(Y,\theta)$ may be taken only in terms of the moments

$\mu_i = E\left[\left\{\frac{\partial t(Y,\theta)}{\partial\theta}\right\}^i\right]$ for i = 2,3,4 and Var $\left[\frac{\partial^2 t(Y,\theta)}{\partial\theta^2}\right]$. We write

$\varepsilon_p = H(\underset{\sim}{\phi}, \underset{\sim}{X}, \underset{\sim}{\theta})$.

The ε_p formulae is, in general, strongly dependent on the asymptotic covariance matrix $(\underset{\sim}{X}^T\underset{\sim}{\phi}\underset{\sim}{W}\underset{\sim}{X})^{-1}$ and therefore on the experimental design used. Some simplification in (2.7) may be obtained by considering some special model matrices.

We assume the model (1.1) and let H_i: $\underset{\sim}{\theta} = F(\underset{\sim}{X}_i \underset{\sim}{\beta}_i)$ i = 1,2 be, respectively, two nested hypotheses of fixed dimensions p_1 and $p_2 (p_2 > p_1)$, where $\underset{\sim}{X}_1$ and $\underset{\sim}{X}_2$ are nxp_1 and nxp_2 known model matrices of ranks p_1 and p_2, $\underset{\sim}{\beta}_1 = (\beta_1 \ldots \beta_{p_1})^T$ and $\underset{\sim}{\beta}_2 = (\beta_1 \ldots \beta_{p_1} \beta_{p_1+1} \ldots \beta_{p_2})^T$. The likelihood ratio criterion for testing H_1 against H_2 is $-2 \log \lambda$ $= 2\left[L(\hat{\underset{\sim}{\beta}}_2) - L(\hat{\underset{\sim}{\beta}}_1)\right]$ where $L(\hat{\underset{\sim}{\beta}}_i)$ is the maximized value of $L(\underset{\sim}{\beta})$ under H_i. The mean of $-2 \log \lambda$ to order n^{-1} can now be calculated $E(-2 \log \lambda) = p_2 - p_1 + H(\underset{\sim}{\phi}, \underset{\sim}{X}_2, \underset{\sim}{\theta}) - H(\underset{\sim}{\phi}, \underset{\sim}{X}_1, \underset{\sim}{\theta})$, where $\underset{\sim}{\theta}$ is the unknown true parameter, and the Bartlett adjustment to test H_1 obtained from $c = 1 + \left[H(\underset{\sim}{\phi},\underset{\sim}{X}_2,\underset{\sim}{\theta}) - H(\underset{\sim}{\phi},\underset{\sim}{X}_1,\underset{\sim}{\theta})\right]/(p_2 - p_1)$. The modified statistic $-2c^{-1} \log \lambda$ under H_1 has the $\chi^2_{p_2-p_1}$ distribution with error $O(n^{-2})$ (Lawley, 1956). This result requires the dimensions of the parameter space under both hypotheses be fixed while the number of observations n becomes large. For this reason, one can not take the hypothesis H_2 as the saturated model.

The improved test of H_1 against H_2 compares $-2\hat{c}^{-1} \log \lambda$, where

$\hat{c} = 1 + \left[H(\underset{\sim}{\phi},\underset{\sim}{X}_2, F(\underset{\sim}{X}_1 \hat{\underset{\sim}{\beta}}_1)) - H(\underset{\sim}{\phi},\underset{\sim}{X}_1, F(\underset{\sim}{X}_1 \hat{\underset{\sim}{\beta}}_1))\right] /(p_2-p_1)$ is the

estimated adjustment under H_1, with the asymptotic $\chi^2_{p_2-p_1}$ distribution.

If the model (1.1) is used with $\phi_\ell = \phi$, $\ell=1\ldots n$ where ϕ is unknown, we have to calculate further terms in (2.1) involving combinations of ϕ and β parameters to get the n^{-1} term in the null expected likelihood ratio statistic.

3. SOME SPECIAL CASES

If we take $t(y,\theta) = y\theta - b(\theta)$, (1.1) is a one parameter exponencial family indexed by the canonical parameter θ and then Cordeiro's (1983) ε_p formulae follows directly from (2.7). In (1.1) we now assume that $t(y,\theta)$ involves a known constant parameter c for all observations, $t(y,\theta) = t(y,\theta,c)$ say, and that $p(\phi) = \phi = 1$ and $q(\phi,y) = q(c,y)$. For doing this several models can be defined within the framework of the extended generalized linear model including: normal-$N(\theta,c^2\theta^2)$, lognormal - $LN(\theta,c^2\theta^2)$, inverse Gaussian $-N^-(\theta,c^2\theta^2)$ distributions with mean θ and known constant coefficient of variation c, gamma $-G(\theta,c)$ distribution with mean θ and known constant scale parameter c and Weibull - $W(\theta,c)$ distribution with mean θ and known constant shape parameter c. Here the normal, gamma and inverse Gaussian distributions are not standard generalized linear models since, in each case, we consider a different parametrization. Using a link function $F^{-1}(\cdot)$ such that $\underset{\sim}{W}$ does not depend on $\underset{\sim}{\theta}$, the information matrix $\underset{\sim}{K}$ is constant and then at least asymptotically, the "covariance stabilizing link" is given by $\eta = k\int \{-D_2(\theta)\}^{1/2} d\theta$ where k is any constant. In particular for the one parameter exponential family $t(y;\theta) = y\theta - b(\theta)$ the covariance stabilizing link comes from $\eta = k\int\{b''(\theta)\}^{1/2} d\theta$. For the models $N(\theta,c^2\theta^2)$, $N^-(\theta,c^2\theta^2)$, $LN(\theta,c^2\theta^2)$ and $W(\theta,c)$ we have $D_2(\theta) = -k_2\theta^{-2}$, $D_3(\theta) = k_3\theta^{-3}$ and $D_4(\theta) = -k_4\theta^{-4}$ where k_2, k_3 and k_4 are known positive functions of c. For these cases the logarithmic is the covariance stabilizing link. If we take this link function $w = g = k_2$, $e = -2k_2$, $f = k_2 - k_3$, $h = -7k_2 + 6k_3 - k_4$, $\underset{\sim}{W} = k_2\underset{\sim}{I}$, $\underset{\sim}{Z} = k_2^{-1}\underset{\sim}{M}$ where $\underset{\sim}{I}$ is the identity matrix of order n and $\underset{\sim}{M} = \{m_{ij}\} = \underset{\sim}{X}(\underset{\sim}{X}^T\underset{\sim}{X})^{-1}\underset{\sim}{X}^T$ represents the orthogonal projection of the sample space onto the subspace spanned by the columns of $\underset{\sim}{X}$. Then the n^{-1} term ε_p reduces to

$$\varepsilon_p = \frac{(-7k_2+6k_3-k_4)}{4k_2^2}\, Tr(\underset{\sim}{M}_d^2) + \frac{(3k_2-k_3)^2}{12k_2^3}\left[2\ \underset{\sim}{1}^T\underset{\sim}{M}^{(3)}\underset{\sim}{1} + 3\ \underset{\sim}{1}^T\underset{\sim}{M}_d\underset{\sim}{M}\underset{\sim}{M}_d\underset{\sim}{1}\right] . \qquad (3.1)$$

For the $N(\theta,c^2\theta^2)$, $N^-(\theta,c^2\theta^2)$, $LN(\theta,c^2\theta^2)$ and $W(\theta,c)$ distribution we get respectively: $k_2 = c^{-2}(1+2c^2)$, $0.5c^{-2}(1+2c^2)$, $[\log(1+c^2)]^{-1}$, c^2, $k_3 = c^{-2}(6+10c^2)$, $c^{-2}(3+c^2)$, $3[\log(1+c^2)]^{-1}$, $c^2(c+3)$ and $k_4 = c^{-2}(36+54c^2)$, $c^{-2}(12+3c^2)$, $11[\log(1+c^2)]^{-1}$, $c^2(c^2+6c+11)$. For the $LN(\theta,c^2\theta^2)$ distribution $-7k_2+6k_3-k_4 = 3k_2-k_3 = 0$ and then by (3.1) $\varepsilon_p = 0$. In this case with log link $L(\underset{\sim}{\beta}) = -\sum_{\ell=1}^{n} \{\log(y_\ell/\hat{\theta}_\ell) + \frac{1}{2}\log(1+c^2)\}^2/\{2\log(1+c^2)\}$ where $\log\theta_\ell = \sum_{i=1}^{p} x_{\ell i}\beta_i$, $\ell = 1 \ldots n$, $\underset{\sim}{\hat{\beta}}$ has a closed form expression $\underset{\sim}{\hat{\beta}} = (\underset{\sim}{X}^T\underset{\sim}{X})^{-1}\underset{\sim}{X}^T[\frac{1}{2}\log(1+c^2)\ \underset{\sim}{1} + \underset{\sim}{v}]$ where $\underset{\sim}{v} = (\log y_1 \ldots \log y_n)^T$ and $cov(\underset{\sim}{\hat{\beta}}) = \log(1+c^2)\ (\underset{\sim}{X}^T\underset{\sim}{X})^{-1}$. Note that the likelihood ratio statistic is given by $2[L(\underset{\sim}{\hat{\beta}}) - L] = \sum_{\ell=1}^{n} [\{\log(y_\ell/\theta_\ell) + \frac{1}{2}\log(1+c^2)\}^2 - \{\log(y_\ell/\hat{\theta}_\ell) + \frac{1}{2}\log(1+c^2)\}^2]/\log(1+c^2)$, and it is distributed exactly as a χ_p^2 in agreement with the result $\varepsilon_p = 0$. For the models $N(\theta,c^2\theta^2)$, $N^-(\theta,c^2\theta^2)$, $LN(\theta,c^2\theta^2)$ and $W(\theta,c)$ with identity link $\underset{\sim}{W} = k_2\underset{\sim}{D}^{-2}$, $\underset{\sim}{E} = -2k_2\underset{\sim}{D}^{-3}$, $\underset{\sim}{G} = \underset{\sim}{0}$, $\underset{\sim}{F} = -k_3\underset{\sim}{D}^{-3}$ where $\underset{\sim}{D} = \mathrm{diag}\{\theta_1 \ldots \theta_n\}$ and $\underset{\sim}{Z} = k_2^{-1}\underset{\sim}{M}$ with $\underset{\sim}{M} = \underset{\sim}{X}(\underset{\sim}{X}^T\underset{\sim}{D}^{-2}\underset{\sim}{X})^{-1}\underset{\sim}{X}^T$ and we get

$$\varepsilon_p = \frac{(12k_3-24k_2-k_4)}{4k_2^2}\, Tr\ (\underset{\sim}{D}^{-4}\underset{\sim}{M}_d^2) + \frac{(k_3^2-12k_2k_3+24k_2^2)}{6k_2^3}\ \underset{\sim}{1}^T\underset{\sim}{D}^{-3}\underset{\sim}{M}^{(3)}\underset{\sim}{D}^{-3}\underset{\sim}{1} +$$

$$\frac{(k_3^2-8k_2k_3+16k_2^2)}{4k_2^3}\ \underset{\sim}{1}^T\underset{\sim}{D}^{-3}\underset{\sim}{M}_d\underset{\sim}{M}\underset{\sim}{M}_d\underset{\sim}{D}^{-3}\underset{\sim}{1}. \qquad (3.2)$$

The fitting of n linearly independent parameters leads to a saturated model with maximum likelihood equations $t'(y_\ell,\hat{\tilde{\theta}}_\ell) = 0$, $\ell = 1 \ldots n$. To test the adequacy of an extended generalized linear model with p linearly independent parameters, the deviance $S_p = 2\sum_{\ell=1}^{n} p(\phi_\ell)[t(y_\ell,\hat{\hat{\theta}}_\ell) - t(y_\ell,\hat{\theta}_\ell)]$ is compared with the χ^2_{n-p} distribution. However, the deviance under the model is not in general distributed as a χ^2_{n-p}, even asymptotically. It is, therefore, essential to have a good approximation to the distribution of S_p if the model under investigation is true. Let $\hat{L}_n$ be the maximized log likelihood for the saturated model. The deviance at the true parameter becomes $2(\hat{L}_n - L) = 2\sum_{\ell=1}^{n} p(\phi_\ell)[t(y_\ell,\hat{\hat{\theta}}_\ell) - t(y_\ell,\theta_\ell)]$ and its expected value can be computed from the distribution of the observations. Then, we have $E(S_p) = 2\,E(\hat{L}_n - L) - (p + \varepsilon_p) + O(n^{-2})$. It may be defined a modified deviance $S^*_p = \frac{n(n-p)}{\hat{E}(S_p)} S_p$ where $\hat{E}(S_p)$ is the maximum likelihood estimate of $E(S_p)$ under the model, such that, with large samples at least, S^*_p is better approximated by a χ^2_{n-p} than is S_p. We now give some special cases for $2\,E(\hat{L}_n - L)$. When $Y\sim N(\theta,c^2\theta^2)$, $\hat{\theta} = y\,\alpha$ with $\alpha = \{(1+4c^2)^{1/2} - 1\}/2c^2$ and $2\,E(\hat{L}_n-L) = n(1-\alpha^2c^2-\log\alpha^2) + \sum_{\ell=1}^{n}[2\log\theta_\ell - E\{\log Y^2_\ell\}]$. But $E\{\log Y^2\} = \log(\theta^2c^2) + E\{\log\chi'^2_1(c^{-2})\}$ where $\chi'^2_1(c^{-2})$ denotes a non-central chi-squared with 1 degree of freedom and noncentrality parameter c^{-2}. From $E\{\log\chi'^2_1(\lambda)\} =$

$$\sum_{j=0}^{\infty}\frac{(\lambda/2)^j e^{-\lambda/2}}{j!}E\{\log\chi^2_{1+2j}\} = \log 2 + \sum_{j=0}^{\infty}\frac{(\lambda/2)^j e^{-\lambda/2}}{j!}\psi(\tfrac{1}{2}+j)$$

where $\psi(\cdot)$ is the digamma function and $\psi(\frac{1}{2}+j) = -\gamma - 2\log 2 + 2\sum_{k=1}^{j}\frac{1}{2k-1}$, $j \geqslant 1$ where $\gamma = -\psi(1)$ is the Euler's constant, we find

$$2E(\hat{L}_n - L) = n\{1 + \gamma + \log(2/\alpha^2c^2) - \alpha^2c^2 - 2\sum_{j=1}^{\infty}\sum_{k=1}^{j}\frac{c^{-2j}\exp(-c^2/2)}{2^j\, j!\,(2k-1)}\}.$$

Introducing $E\{\log Y^2\} = 2\log\theta - c^2 + O(\theta^{-4})$, $2E(\hat{L}_n - L)$ is approximately given by $n[\alpha + c^2 - \log\alpha^2]$. For $Y\sim N^-(\theta,c^2\theta^2)$, $\hat{\hat{\theta}} = y\alpha$ with $\alpha = \{c^2 + (c^2+4)^{1/2}\}/2$ and $2E(\hat{L}_n - L) = \frac{2n}{\alpha+1} + n\log\alpha + \sum_{\ell=1}^{n}\{E(\log Y_\ell) - \log\theta_\ell\}$. From $E(\log Y) \doteq \log\theta_\ell - \frac{c^2}{2}$ we get $2E(\hat{L}_n - L) \doteq n[\frac{2}{\alpha+1} + \log\alpha - \frac{c^2}{2}]$. If $Y\sim LN(\theta,c^2\theta^2)$, $\hat{\hat{\theta}} = y(1+c^2)^{1/2}$ and it is easily found $2E(\hat{L}_n - L) = n$. Considering $Y\sim W(\theta,c)$, $\hat{\hat{\theta}} = y\,\Gamma(1+c^{-1})$ and we have $2E(\hat{L}_n - L) = 2n\gamma \doteq 1.1544n$.

For $Y\sim G(\theta,c)$, $\psi(c\hat{\hat{\theta}}) = \log(cy)$. If c is large enough the approximation $\psi(c\,\hat{\hat{\theta}}) \doteq \log(c\,\hat{\hat{\theta}} - \frac{1}{2})$ may be used and then $\hat{\hat{\theta}} \doteq y + \frac{1}{2c}$. Using also the approximation $\log\Gamma(x) \doteq (x-\frac{1}{2})\log(x-\frac{1}{2}) - (x-\frac{1}{2})$ to calculate $\log\Gamma(cy+\frac{1}{2})$ and $\log\Gamma(c\theta)$ for c large we can get after some

algebra $2E(\hat{L}_n - L) \doteq n$.
Supposing that $t(y,\theta) = y\theta - b(\theta)$ we have $\hat{\hat{\theta}} = b'^{-1}(y)$ and the expectation $2E(\hat{L}_n - L) = 2 \sum_{\ell=1}^{n} [E\{Y\, b'^{-1}(Y) - b(b'^{-1}(Y))\} - \theta\, b'(\theta) + b(\theta)]$ is discussed in Cordeiro (1983) for several generalized linear models. For the log-gamma distribution with known constant variance and density $\Pi(y;\theta,\phi) = c(\phi) \exp [\phi\{y - \theta - \exp(y-\theta)\}]$, $y > 0$, where $c(\phi)$ is a norming constant we have $\hat{\hat{\theta}} = y$, $E(Y) = \theta + 1 - \frac{\partial \log c(\phi)}{\partial \phi}$ and then $2E(\hat{L}_n - L) = 2n\phi \frac{\partial \log c(\phi)}{\partial \phi}$. Considering the positive hyperbolic distribution with density $\Pi(y;\theta) = \frac{y^{-1}}{2\, K_0(1)} \exp [- \frac{1}{2}\{y^{-1} \exp(\theta) + y \exp(-\theta)\}]$, where $K_\nu(\cdot)$ is the Bessel function with index ν, $E(Y) = \frac{K_1(1)}{K_0(1)} \exp(\theta)$, $\hat{\hat{\theta}} = \log y$ and we find $2\, E(\hat{L}_n - L) = 2n[\frac{K_1(1)}{K_0(1)} - 1]$ $\doteq 0.8592\, n$.

4. THE 1 - WAY CLASSIFICATION MODEL

We consider p independent random samples of sizes $n_1 \ldots n_p$ ($n_i \geq 1$, $i = 1 \ldots p$) taken from p populations with distribution function (1.1) indexed by the parameters θ_i, ϕ_i. The vector of observations is $\underset{\sim}{y} = (y_{11} \ldots y_{1n_1} \ldots y_{p1} \ldots y_{pn_p})$ and the linear structure of the model is taken as $\eta_i = F^{-1}(\theta_i) = \beta + \beta_i$, $i = 1 \ldots p$, with $\beta_+ = 0$, where β is a general mean and β_i is the effect on the response of the i - th population. A typical element of $\underset{\sim}{Z}$ is $\delta_{ij} [n_i\, p(\phi)\, w_i]^{-1}$ where $\delta_{ij} = 1$ if i and j index observations in the same population, and zero otherwise. After some algebra we can get the n^{-1} term in the null expected deviance as

$$\varepsilon_p = \sum_{i=1}^{p} [12\, D_{2i}\, (D''_{2i} - D'_{3i}) + 3\, D_{2i}\, D_{4i} + 24\, D'_2\, D_{3i} - 5\, D^2_{3i} - 24\, D'^2_{2i}] / [12\, n_i\, p(\phi_i)\, D^3_{2i}] \tag{4.1}$$

where $D_{2i} = D_2(\theta_i)$, $D'_{2i} = \frac{\partial D_2(\theta_i)}{\partial \theta_i}$ and so on. Note that (4.1) is obtained without any link specification and, as is to be expected, ε_p is not a function of the link since here $\hat{\theta}_i$ comes from $\sum_{j=1}^{n_i} t'(y_{ij}, \hat{\theta}_i) = 0$ and does not depend on the link. For the models $N(\theta, c^2\theta^2)$, $N^-(\theta, c^2\theta^2)$, $W(\theta, c)$ and $LN(\theta, c^2\theta^2)$ discussed in section 3, (4.1) reduces to

$$\varepsilon_p = (24\, k_2^2 + 5\, k_3^2 - 12\, k_2\, k_3 - 3\, k_2\, k_4)\, (\sum_{i=1}^{p} n_i^{-1}) / 12\, k_2^3. \tag{4.2}$$

For $Y \sim LN(\theta, c^2\theta^2)$, $\varepsilon_p = 0$ as it is expected since $2\{L(\hat{\beta})-L\} = [\log(1+c^2)]^{-1} \sum_{i=1}^{p} n_i [\log(\tilde{y}_i/\theta_i) + \frac{1}{2}\log(1+c^2)]^2$ is χ^2_p where $\tilde{y}_i$ is the geometric mean of $y_{i1} \ldots y_{in_i}$. For the gamma, log-gamma and positive hyperbolic distribution discussed earlier we find, respectively,

$$\varepsilon_p = \frac{1}{12} \sum_{i=1}^{p} [5\,\psi''(c\theta_i)^2 - 3\,\psi'(c\theta_i)\,\psi'''(c\theta_i)] / [n_i\,\psi'(c\theta_i)^3],$$

$$\varepsilon_p = \frac{1}{6} \sum_{i=1}^{p} n_i^{-1}\phi_i^{-1} \text{ and } \varepsilon_p = \frac{-K_0(1)}{4\,K_1(1)} \sum_{i=1}^{p} n_i^{-1}.$$

REFERENCES

Cordeiro G M (1983) Improved likelihood ratio statistics for generalized linear models. JRSS-B, 45, 3, 404-413.

Jørgensen B (1983) Maximum likelihood estimation and large-sample inference for generalized linear and nonlinear models. Biometrika, 70, 1, 19-28.

Lawley D N (1956) A general method for approximating to the distribution of likelihood ratio criteria. Biometrika, 43, 295-303.

Nelder J A and Wedderburn R W M (1972) Generalized linear models. JRSS-A 135, 370-384.

COMPARING ESTIMATED SPECTRAL DENSITIES USING GLIM

Peter J. Diggle
CSIRO Division of Mathematics and Statistics
Canberra, ACT 2601 Australia

SUMMARY

A test due to Coates and Diggle (1985) for comparing two estimated spectra is re-formulated as a generalised linear model with gamma errors and logarithmic link. This provides an immediate extension to the case of $p > 2$ spectra. GLIM macros for implementing the test are listed. An application to some medical data concerning natural fluctuations in the level of luteinising hormone in blood-samples is described.

Keywords: time-series, spectral analysis, luteinising hormone.

1. INTRODUCTION

Suppose that two time-series $\{x_{1t}: t=1,\ldots,n\}$ and $\{x_{2t}: t=1,\ldots,n\}$ are generated as independent realisations of stationary processes $\{X_{1t}\}$ and $\{X_{2t}\}$, and that we wish to test the hypothesis that the series are in fact generated from the same process, $\{X_t\}$ say. Coates and Diggle (1985) develop a test, based on the spectral properties of the two series, which makes no parametric assumptions about $\{X_t\}$. In this note, we re-formulate their test using the framework of the generalised linear model. The re-formulation provides an immediate extension to the case of $p > 2$ series $\{X_{jt}\}$, and a convenient means of implementation via GLIM. We give GLIM macros for this purpose and describe an applicaion to some medical data. For a general discussion of the spectral analysis of time-series data, see Priestley (1981) or, at a more elementary level, Chatfield (1980, Ch.6,7).

2. FORMULATION AND RE-FORMULATION

(2.1) Definitions and basic distributional properties

The spectrum of a stationary process $\{X_t\}$ is the discrete Fourier transform of its autocovariance function. Thus, if $\gamma_k = \mathrm{Cov}\{X_t, X_{t-k}\}$, the spectrum is defined to be the real-valued function

$$f(\omega) = \sum_{k=-\infty}^{\infty} \gamma_k e^{-ik\omega} \qquad (0 < \omega < \pi) .$$

The periodogram of a time-series $\{x_t : t=1,\ldots,n\}$ is defined as

$$I(\omega) = n^{-1}[\{\sum_{t=1}^{n} x_t \cos(t\omega)\}^2 + \{\sum_{t=1}^{n} x_t \sin(t\omega)\}^2].$$

The periodogram is usually computed at a discrete set of values of ω, the Fourier frequencies $\omega_k = 2\pi k/n : k=1,\ldots,m=[(n-1)/2]$. The asymptotic distribution theory of the periodogram ordinates $I(\omega_k)$ is remarkably simple, and can be summarised by the following:

(i) $I(\omega) \sim f(\omega)\ \chi_2^2/2$ (2.1)

(ii) for $k' \neq k$, $I(\omega_k)$ and $I(\omega_{k'})$ are independent. (2.2)

These results involve mild assumptions about the process $\{X_t\}$; for precise statements and proofs, see Priestley (1981, Section 6.2.2), but note the difference of a factor 2π in the definition of $f(\omega)$.

2.2. The test proposed by Coates and Diggle

Let $f_j(\omega)$ and $I_j(\omega)$ denote the spectrum of $\{X_{jt}\}$ and periodogram of $\{x_{jt}\}$ respectively. Define $\mu_k = \ln\{f_1(\omega_k)/f_2(\omega_k)\}$ and $Y_k = \ln\{I_1(\omega_k)/I_2(\omega_k)\}$, $k=1,\ldots,m$. It follows from (2.1) and (2.2) that the Y_k are independent logistic variates with location parameters μ_k, i.e. the distribution function of Y_k is

$$F_k(y) = \{1+\exp(\mu_k - y)\}^{-1}$$

Coates and Diggle (1985) postulate that

$$\mu_k = \alpha + \beta k + \gamma k^2 \qquad (2.3)$$

and use generalised likelihood ratio methods to test the hypotheses

$$H_{00} : \alpha = \beta = \gamma = 0 ,$$

$$H_0 : \beta = \gamma = 0 .$$

Note that H_{00} corresponds to the stated null hypothesis of a common underlying process $\{X_t\}$, except that it allows a possible difference in location, $E(X_{1t}) \neq E(X_{2t})$, whereas H_0 additionally allows a difference in scale, $Var(X_{1t}) \neq Var(X_{2t})$.

The quadratic relationship (2.3) is not proposed as a serious scientific model. Rather, it is merely a convenient way of building into the analysis a degree of smoothness which is to be expected in practice, given the natural ordering of the frequency scale. Note also that (2.3) involves no parametric assumptions about the $f_j(\omega)$ themselves. Coates and Diggle (1985) found in Monte Carlo studies that (2.3) gave improved power characteristics by comparison with, on the one hand, a simpler, linear formulation for the μ_k and on the other a fully non-parametric approach which does not assume any functional relationship amongst the μ_k. In this sense, (2.3) represents a balance between flexibility and persimony. Clearly, other parameterisations are possible, and might perform at least as well. We do not pursue this issue here.

(2.3) A re-formulation

Coates and Diggle's formulation in terms of log periodogram ratios proved convenient for the development of a generalised likelihood ratio test from first principles, not least because the similarity of the logistic distribution to the Normal meant that ordinary least squares estimation provided excellent first approximations to the maximum likelihood estimates of α, β, γ. Within the GLIM framework, it is more natural to work directly with the periodogram ordinates, whose sampling distributions are proportional to chi-squared on two degrees of freedom.

Suppose that p time-series $\{x_{jt}: t=1,\ldots,n\}$ are generated as independent realisations of stationary processes $\{X_{jt}\}$, for each of $j=1,\ldots,p$. Let $f_j(\omega_k)$ and $I_j(\omega_k)$ be the corresponding spectra and periodograms evaluated at the Fourier frequencies ω_k. Define random variables $Y_{kj}= I_j(\omega_k)$ and parameters $\theta_{kj}=f_j(\omega_k)$. Then, according to (2.1) and (2.2) the Y_{kj} are asymptotically independent exponential variates with expectations θ_{kj}. In GLIM terminology, this defines a saturated generalised linear model with exponential error distribution. Now, postulate a log-linear model with

$$\ln \theta_{kj} = \tau_k + \alpha_j + \beta_j k + \gamma_j k^2 \quad . \tag{2.4}$$

In the case p=2, (2.4) is equivalent to (2.3). More generally, the hypothesis of a common underlying process corresponds to the sub-model

$$\ln \theta_{kj} = \tau_k , \tag{2.5}$$

whilst possible scale-differences amongst the $\{X_{jt}\}$ are accommodated by

$$\ln \theta_{kj} = \tau_k + \alpha_j . \tag{2.6}$$

The parameters β_j and γ_j in (2.4) again represent a smooth parameterisation of differences amongst the shapes of the spectra of $\{X_{jt}\}$ whilst the τ_k are nuisance parameters corresponding to an unspecified "base-spectrum" $f(\omega_k):k=1,\ldots,m$. Generalised likelihood ratio statistics can be computed from a table of deviances associated with with the models (2.4), (2.5) and (2.6).

These tests are intended to be sensitive to "smooth" departures from identical or proportional spectra, for example a shift in power towards the lower or higher end of the frequency range, or a general flattening or sharpening of a single peak. To this extent at least, the fitted linear and quadratic parameters β_j and γ_j admit a qualitative interpretation. For a quantitative comparison, direct inspection of smoothed periodograms or periodograms ratios, together with pointwise confidence limits for the underlying spectra or spectral ratios, seems preferable on grounds of flexibility.

3. GLIM MACROS

The macro PERIODOGRAM computes the periodogram of a single time-series. Its four formal arguments identify the time-series itself, the length of the series, the corresponding periodogram and the number of periodogram ordinates. The first two arguments must be assigned values before the macro is called, whilst the remaining two are assigned values during execution. The macro PLOOP is called repeatedly from PERIODOGRAM, and its invocation therefore requires no action by the user. Finally, the macro COMPARE produces a sequence of scaled deviances from fits corresponding to the models (2.5), (2.6) and (2.4) respectively. Prior to execution, the user must assign values to scalars p, the number of periodograms to be compared, and m, the number of ordinates of each periodogram. He must also assign to a vector, y say, of length mp the concatenation of all p periodograms. The three formal arguments for COMPARE are then y, p and m respectively.

Listings of these three macros are given in the appendix. For long series it would be preferable to compute the periodograms externally using a fast Fourier transform algorithm. Monro (1976) gives a FORTRAN algorithm suitable for series of length n a power of 2, whilst both the NAG and IMSL subroutine libraries implement a method due to Singleton (1967) which copes with arbitrary n.

4. AN APPLICATION : FREQUENCY OF PULSATION OF LUTEINISING HORMONE

Luteinising hormone (LH) is known to display a pulsatile release pattern. Determination of the frequency of pulsation in normal and abnormal conditions has potential implications for diagnosis and treatment. Examination of time-series of LH levels suggests an imperfect cyclic pattern which should be amenable to spectral analysis. Murdoch et al (1985) show periodograms of individual series from normal women, each sampled at five minute intervals over a period of several hours, and identify various spectral features.

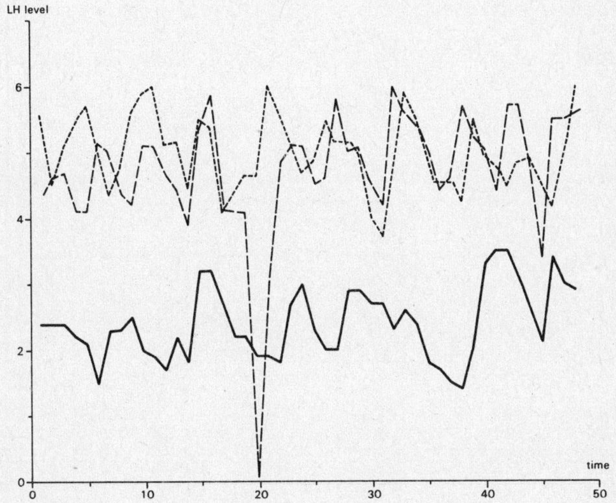

Figure (1) . Three time-series of LH levels
—early follicular phase; ...late follicular phase (first cycle)
---late follicular phase (second cycle)

One general question which arises is the extent to which these spectral features are reproducible from replicate series: extension of indiviudal series is impractical because each determination of LH involves taking a blood sample. Figure (1) shows three series obtained from one subject. In each case, LH levels have been determined every 10 minutes over an eight-hour period. One of the series relates to the early follicular phase of the subject's menstrual cycle, the remaining two to the late follicular phase of two successive cycles.

One of the late-phase series contains a gross outlier: for the subsequent analyses we replace this by the sample mean of the remaining 47 observations. The generally lower LH values in the early phase are well known. Recall that spectral properties are unaffected by a location shift.

Table (1). Analysis of deviance for LH data

(a) all three series

Model	Deviance	df
τ_k	46.42	46
$\tau_k + \alpha_j$	43.48	44
$\tau_k + \alpha_j + \beta_j k + \gamma_j k^2$	30.64	40

(b) two late follicular series

Model	Deviance	df
τ_k	21.85	23
$\tau_k + \alpha_j$	21.02	22
$\tau_k + \alpha_j + \beta_j k + \gamma_j k^2$	18.21	20

Table (1) gives deviances associated with models (2.4) to (2.6), applied to the periodograms of

(a) all three series

(b) two late-phase series only

Table (1a) gives strong evidence for differences in shape amongst the three spectra ($\chi_4^2 = 12.84$, $p \simeq 0.012$), whilst Table (1b) gives no evidence against the hypothesis of a common spectrum for the two late follicular series.

These results suggest combining the periodograms from the two late-phase series to produce a single estimate of the underlying spectrum. To do this, we first average the two sets of periodogram ordinates at each of the 23 Fourier frequencies, then take an unweighted moving average over three successive Fourier frequencies to smooth out sampling fluctuations. Figure (2a) shows the resulting spectral estimates, together with point-wise 90% confidence limits based on the observation that the approximate sampling distribution of each spectral estimate is proportional to χ_{12}^2

(except at the end-points). The estimated spectral density shows a moderately substantial peak around the ninth Fourier frequency, corresponding to a cycle of length 53 minutes approximately.

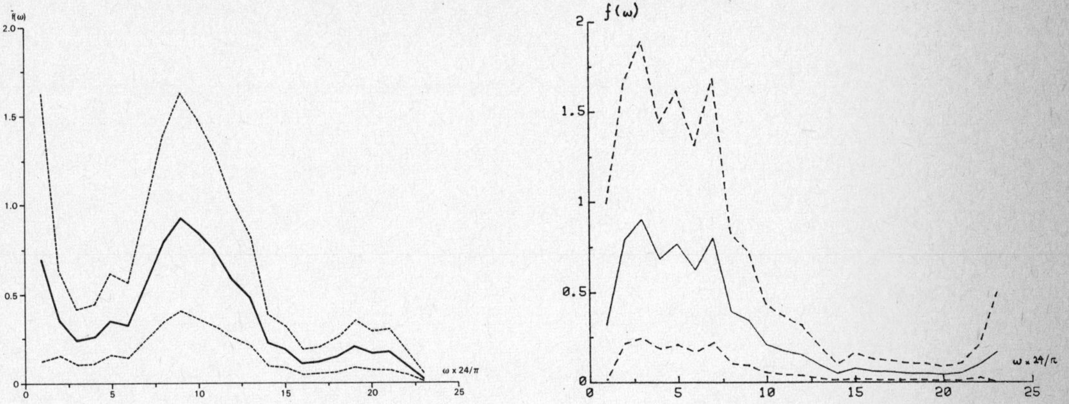

(a) late follicular phase

(b) early follicular phase

Figure (2). Spectral estimates.

______point estimates; ....90% confidence limits

Other weaker spectral features are a concentration of power towards low frequencies, and a very small, flat peak around the twentieth Fourier frequency. The former is typical of positively autocorrelated time-series, but may also reflect diurnal variation or other long-term changes in mean LH levels. The latter implies an element of rapid oscillation in LH levels which, whilst intriguing, is of doubtful clinical importance. Qualitatively similar spectral features were reported by Murdoch et al (1985).

For comparison, Figure (2b) shows spectral estimates for the early-phase series obtained by applying a three-point moving average to the periodogram, together with point-wise 90% confidence limits. With only six, rather than 12, degrees of freedom for each estimate, spectral features are less well defined. The general impression is of a concentration of power at low frequencies, with a hint of a broad peak covering the third to seventh Fourier frequencies.

ACKNOWLEDGEMENT

The ideas in Section (2.3) of the paper arose out of discussions with David Williams after a meeting of the Edinburgh Local Group of the Royal Statistical Society in November 1984.

REFERENCES

Chatfield, C. (1980). The Analysis of Time Series : an introduction. London: Chapman and Hall.

Coates, D.S. and Diggle, P.J. (1985). Tests for comparing two estimated spectral densities. J. Time Series Anal. (to appear).

Monro, D.M.L. (1976). Algorithm AS97. Real discrete fast Fourier transform. Appl. Statist. 25, 166-72.

Murdoch, A.P., Diggle, P.J., Dunlop, W. and Kendall-Taylor, P. (1985). Determination of the frequency of pulsatile luteinising hormone secretion by time series analysis. Clin. Endocrin. 22, 341-6.

Priestley, M.B. (1981). Spectral Analysis and Time Series. London: Academic Press.

Singleton, R.C. (1976). On computing the fast Fourier transform. Comm. ACM 10, 647-54.

APPENDIX

```
$macro periodogram
$c
$c    Computation of periodogram of a single time-series.
$c    Four formal arguments represent:
$c          %1 time-series       (input)
$c          %2 length of series (input)
$c          %3 periodogram       (output)
$c          %4 number of periodogram ordinates (output)
$c
$calc %4=%tr((%2-1)/2)$vari %4 %3:%2 tl wtl cl sl
$calc tl=%gl(%2,1) : %i=%4
$arg ploop %1 %2 %3
$while %i ploop$calc %3=%3/%2 $delete tl wtl cl sl
$endmac
$macro ploop
$c
$c   Computes periodogram ordinate for current Fourier frequency
$c   and decrements this frequency by one unit.
$c   Called from macro PERIODOGRAM, and uses same formal arguments.
$c
$calc %w=2*%pi*%i/%2 : %u=%pi/2 : wtl=%w*tl
$calc sl=%sin(wtl) : cl=%sin(wtl+%u)
$calc %3(%i)=(%cu(%1*cl)**2)+(%cu(%1*sl)**2)
$calc %i=%i-1
$endmac
$macro compare
$c
$c    Comparison of 2 or more periodograms via the model in Coates and
$c    Diggle (1985, J. Time Series Anal.). Formal arguments are:
$c          %1 vector containing catenated periodograms (input)
$c          %2 number of periodograms                   (input)
$c          %3 length of each periodogram               (input)
$c    Output is table of deviances from fits corresponding to:
$c          identical spectra
$c          proportional spectra
$c          quadratic log-linear model for differences in shape.
$c
$calc %n=%2*%3$units %n
$calc repl=%gl(%2,%3):fl=%gl(%3,1):%z=%cu(fl)/%nu:linl=fl-%z:quadl=linl*linl
$factor repl %2:fl %3
$yvar %1$error g$scale 1$link 1$fit fl:+repl:+repl.linl+repl.quadl$
$delete repl:fl:linl:quadl
$endmac
$return
```

Semi-parametric Generalized Linear Models

by

Peter J. Green* and Brian S. Yandell

Department of Statistics
University of Wisconsin-Madison

1. Introduction

When the form of a regression relationship with respect to some but not all of the explanatory variables is unknown, the statistician is caught in a quandary. Should parametric models be abandoned altogether, thus losing the opportunity of estimating parameters of real interest and sacrificing efficiency in estimation and prediction, or should the extraneous variables be forced into a parametric model by imposing a possibly inappropriate functional form without adequate justification?

A compromise is possible using the idea of semi-parametric modelling. This has been considered by several authors in varying degrees of generality; see for example Rice (1981), Green, Jennison and Seheult (1983), Wahba (1984), and Green (1985b). In the present context of generalized linear models we consider replacing the familiar linear predictors $\eta_i = \mathbf{x}_i^T \boldsymbol{\beta}$ by the more general predictors

$$\theta_i = \mathbf{x}_i^T \boldsymbol{\beta} + \gamma(\mathbf{t}_i) \tag{1}$$

with $\mathbf{x}_i$ a p-vector of explanatory variables for the ith observation, $\boldsymbol{\beta}$ the corresponding regression coefficients, $\mathbf{t}_i$ the scalar or vector of extraneous variables, and γ a function or curve whose form is not specified. As a simple example, imagine a binomial logistic regression in which the "intercept" term is believed to vary in time or with geographical location. In section 5 we consider in detail an example where a previous investigator has been unsure about the precise dependence of the binary response (presence of tumour) on one of four explanatory variables (age at death).

Straightforward maximization of the log-likelihood function L, which we will write in the composite form $L(\boldsymbol{\theta}(\boldsymbol{\beta},\gamma))$ to emphasize the roles of predictors, parameters, and unknown curve, is no longer appropriate as a method of estimation. This leads to overfitting in the absence of any constraints on $\boldsymbol{\beta}$. Indeed, it typically renders the parameters $\boldsymbol{\beta}$ unidentifiable. But progress is possible by maximizing instead a penalized version of the log-likelihood, if we are willing to place weak constraints on the form of γ by assuming that it is smooth. Thus we maximize the penalized log-likelihood

$$L(\boldsymbol{\theta}(\boldsymbol{\beta},\gamma)) - \tfrac{1}{2}\lambda J(\gamma) \tag{2}$$

where the penalty functional J is some numerical measure of the "roughness" of γ. This might be adopted on ad-hoc grounds (for example, an integrated squared derivative of γ), or follow from a Bayesian argument specifying a prior distribution for γ. The scalar λ is a tuning constant, used to

* present affiliation: Department of Mathematical Sciences, University of Durham.

regulate the smoothness of the fitted curve γ. Typically we try a range of values for λ in an exploratory fashion, as well as considering automatic choice based on the data. One ultimate aim may be to discover the form of γ in the hope of modelling it parametrically in future.

2. Maximum Penalized Likelihood Estimates

Here we will only consider maximization of (2) over γ in the span of a set of q prescribed basis functions $\{ \phi_k ; k = 1,2,\ldots,q \}$: we write

$$\gamma = \sum_{k=1}^{q} \xi_k \phi_k$$

and assume in addition that J satisfies

$$J(\sum_{k=1}^{q} \xi_k \phi_k) = \xi^T \mathbf{K} \xi \tag{3}$$

for some $q \times q$ non-negative definite symmetric $\mathbf{K}$. We thus re-write the penalized log-likelihood in the form

$$L(\boldsymbol{\theta}(\boldsymbol{\beta},\boldsymbol{\xi})) - \tfrac{1}{2}\lambda \boldsymbol{\xi}^T \mathbf{K} \boldsymbol{\xi} \tag{4}$$

to be maximized over choice of the vectors $\boldsymbol{\beta}$ and $\boldsymbol{\xi}$.

This finite-dimensional approach is not intended to compromise our non-parametric assumptions about the curve γ. The dimension q, perhaps equal to n, will typically be too large for parametric estimation of ξ to be appropriate, and the basis functions will be chosen so as not to materially constrain the curve, except perhaps in fine detail. With certain penalty functionals, for example those used in spline smoothing, it turns out that with $q = n$ we are not imposing any constraints at all (see section 3).

The semi-parametric regression problem expressed in the general form (4) is considered in some detail by Green (1985b), who derives the following iterative scheme for the maximum penalized likelihood estimates (MPLEs) $\hat{\boldsymbol{\beta}}$ and $\hat{\boldsymbol{\xi}}$. Suppose we have trial estimates $\boldsymbol{\beta}$ and $\boldsymbol{\xi}$. Using these, compute the n-vector of scores $\mathbf{u}$ and the $n \times n$ information matrix $\mathbf{A}$:

$$\mathbf{u} = \frac{\partial L}{\partial \boldsymbol{\theta}}, \qquad \mathbf{A} = E\left[-\frac{\partial^2 L}{\partial \boldsymbol{\theta} \partial \boldsymbol{\theta}^T} \right].$$

In the case of a generalized linear model, $\mathbf{u}$ and $\mathbf{A}$ may be expressed as

$$u_i = \frac{\pi_i (y_i - \mu_i)}{\phi \tau_i^2 \delta_i}, \qquad \mathbf{A} = diag\left[\frac{\pi_i}{\phi \tau_i^2 \delta_i^2} \right],$$

in the notation of the GLIM3 manual (Baker and Nelder, 1978), where the vectors $\boldsymbol{\mu}$, $\boldsymbol{\tau}^2$, and $\boldsymbol{\delta}$ are computed with $\boldsymbol{\theta}$ replacing the linear predictor. We also need the $n \times p$ and $n \times q$ matrices of derivatives

$$\mathbf{D} = \frac{\partial \boldsymbol{\theta}}{\partial \boldsymbol{\beta}}, \qquad \mathbf{E} = \frac{\partial \boldsymbol{\theta}}{\partial \boldsymbol{\xi}}.$$

Then updated estimates $(\boldsymbol{\beta}^*, \boldsymbol{\xi}^*)$ are obtained as the solution to the linear system

$$\begin{bmatrix} \mathbf{D}^T\mathbf{A}\mathbf{D} & \mathbf{D}^T\mathbf{A}\mathbf{E} \\ \mathbf{E}^T\mathbf{A}\mathbf{D} & \mathbf{E}^T\mathbf{A}\mathbf{E} + \lambda\mathbf{K} \end{bmatrix} \begin{bmatrix} \beta^* \\ \xi^* \end{bmatrix} = \begin{bmatrix} \mathbf{D}^T \\ \mathbf{E}^T \end{bmatrix} \mathbf{A}\mathbf{Y}, \tag{5}$$

where

$$\mathbf{Y} = \mathbf{A}^{-1}\mathbf{u} + \mathbf{D}\beta + \mathbf{E}\xi.$$

This scheme is based on the Newton-Raphson method with Fisher scoring, and these updating equations can be seen to combine the iteratively reweighted least squares equations for β used in GLIM (see also Green (1984)), with ridge-regression type equations for ξ (O'Sullivan, Yandell and Raynor, Jr., 1984; Yandell, 1985).

As they stand, the equations (5) are not ideal for practical computation. It is the purpose of this paper to derive various algorithms implementing this scheme, and to illustrate semi-parametric modelling applied to both real and simulated data sets.

3. The One-Dimensional Case, with Cubic Spline Smoothing

For a restrictive but very useful special case, consider a generalized linear model with predictors $\{\theta_i\}$ given by (1), in which $\{ t_i \}$ are one-dimensional, and suppose that the roughness penalty $J(\gamma)$ takes the form $\int (\gamma''(t))^2 \, dt$. This allows one additional explanatory variable to enter in a non-parametric fashion; the form of penalty used ensures that the dependence on this variable is "visually smooth". Some aspects of the purely non-parametric version of this problem were discussed by Silverman (1985).

For simplicity, we suppose that the $\{ t_i \}$ are distinct and ordered, $t_1<t_2<...<t_n$, but relaxing this requirement presents no great difficulty. It is well known (Reinsch, 1967) that the γ maximizing (2) for any fixed β and λ is a natural cubic spline with knots at $\{ t_i \}$, that the space of such splines has dimension n, and that we may choose a basis for this space with $\phi_k(t_i) = \delta_{ik}$.

In this case the notation used above simplifies: $q = n$, $\mathbf{D}$ and $\mathbf{E}$ are constant matrices with $\mathbf{D}$ having i th row equal to $\mathbf{x}_i^T$ and $\mathbf{E}$ being the identity, $\mathbf{A}$ is diagonal, and $\mathbf{Y} = \mathbf{A}^{-1}\mathbf{u} + \theta$. Further, it is implicit in Reinsch (1967) that $\mathbf{K}$ may be written $\Delta^T\mathbf{W}^{-1}\Delta$ where Δ is the $(n-2)\times n$ matrix taking second differences:

$$\Delta_{ii} = h_i^{-1}, \ \ \Delta_{ii+1} = -(h_i^{-1} + h_{i+1}^{-1}), \ \ \Delta_{ii+2} = h_{i+1}^{-1}.$$

and $\mathbf{W}$ is the symmetric tridiagonal matrix of order $(n-2)$:

$$W_{i-1i} = W_{ii-1} = h_i/6, \ \ W_{ii} = (h_i + h_{i+1})/3.$$

where $h_i = t_{i+1} - t_i$. The important point about this decomposition is that Δ and $\mathbf{W}$ are banded.

One possible algorithm implementing (5) involves an inner iteration between the pair of equivalent equations

$$\beta^* = (\mathbf{D}^T\mathbf{A}\mathbf{D})^{-1}\mathbf{D}^T\mathbf{A}(\mathbf{Y}-\xi^*)$$

$$\xi^* = \mathbf{S}\,(\mathbf{Y}-\mathbf{D}\beta^*) \tag{6}$$

where $\mathbf{S} = (\mathbf{A} + \lambda\mathbf{K})^{-1}\mathbf{A}$. This will always converge (Green, 1985a). But further iteration can be avoided by eliminating ξ^* from (5) to give

$$\beta^* = (\mathbf{D}^T\mathbf{A}(\mathbf{I}-\mathbf{S})\mathbf{D})^{-1}\mathbf{D}^T\mathbf{A}(\mathbf{I}-\mathbf{S})\mathbf{Y}. \tag{7}$$

Solution of this small $(p \times p)$ system for β^* is followed by use of (6) to obtain ξ^*. From the updated (β^*, ξ^*) we recompute θ, thence $\mathbf{u}$ and $\mathbf{A}$, and the cycle is repeated to convergence.

This approach is highly practicable, and very economical, since apart from solving the linear equations (7), and some matrix multiplications, we only need to apply the "smoothing operator" $\mathbf{S} = (\mathbf{A} + \lambda\mathbf{K})^{-1}\mathbf{A}$ to form $\mathbf{SY}$ and $\mathbf{SD}$. But a consequence of the special structure of $\mathbf{K}$ mentioned above is that $\mathbf{S}$ can be applied to a vector in only $O(n)$ operations. We use a minor modification of the version of Reinsch's algorithm given by De Boor (1978) to obtain a very fast implementation.

An almost identical approach may be adopted more generally, with any penalty functional for which $\mathbf{S} = (\mathbf{A} + \lambda\mathbf{K})^{-1}\mathbf{A}$ may pre-multiply a vector in $O(n)$ time. This could include splines of different orders, penalties based on discrete differences, and "moving average" smoothers.

4. Goodness-of-fit, Standard Errors, and Choice of λ

Goodness-of-fit can be assessed globally, as in generalized linear models, by the deviance $\Delta = 2\{sup_\theta\, L(\theta) - L(\theta(\hat{\beta},\hat{\xi}))\}$ where $(\hat{\beta},\hat{\xi})$ are the MPLEs. Locally, it is measured by residuals: either the deviance residuals, the signed square roots of the individual contributions to Δ, or in GLIM fashion as

$$z_i = u_i \,/\, A_{ii}^{1/2} = \left[\frac{\pi_i}{\phi}\right]^{1/2}\left[\frac{y_i-\mu_i}{\tau_i}\right]. \tag{8}$$

This is all standard, but we do need a new concept of degrees-of-freedom to assign to Δ. It turns out (Green, 1985b) that the appropriate value, not in general an integer, is given by

$$\nu = n - tr(\mathbf{S}) - tr\left[(\mathbf{D}^T\mathbf{A}(\mathbf{I}-\mathbf{S})\mathbf{D})^{-1}\mathbf{D}^T\mathbf{A}(\mathbf{I}-\mathbf{S})^2\mathbf{D}\right]. \tag{9}$$

This is an approximation to the asymptotic expectation of Δ, and reduces to the usual $n - rank(\mathbf{D})$ when the non-parametric part of the model is omitted. In the non-parametric case, this ν has been used informally for linear models (Eubank, 1984; Eubank, 1985) and generalized linear models (O'Sullivan, Yandell and Raynor, Jr., 1984; O'Sullivan, 1985; Yandell, 1985).

Similar somewhat approximate asymptotics lead to an estimated variance matrix for $\hat{\beta}$ of the form

$$(\mathbf{D}^T\mathbf{A}(\mathbf{I}-\mathbf{S})\mathbf{D})^{-1}\mathbf{D}^T\mathbf{A}(\mathbf{I}-\mathbf{S})^2\mathbf{D}(\mathbf{D}^T\mathbf{A}(\mathbf{I}-\mathbf{S})\mathbf{D})^{-1} \tag{10}$$

from which standard errors may be calculated. In the absence of the appropriate distribution theory, neither the deviance nor the standard errors should be used in formal significance tests, at present, but they do seem to provide adequate guidelines for model selection.

Computation of these quantities follows naturally from the algorithm outlined in section 3, and consists of solving $p \times p$ linear systems following the repeated application of $\mathbf{S}$ to $\mathbf{D}$. The only part of this that is not simple to implement in $O(n)$ time is the first trace term, $tr(\mathbf{S})$, which in our present program takes about $7n^2$ multiplications or divisions. However an $O(n)$ algorithm for this computation in linear spline smoothing has recently been announced by O'Sullivan (1985), and we will adapt this to the present context.

As for automatic choice of λ , Wahba's generalized cross-validation (GCV) method (Wahba, 1977), which uses an invariant modification of a predictive mean-squared error criterion, may be adapted to this situation. A quadratic approximation to the quantity to be minimized (over λ) is simply Δ/v^2 , so no further computation is involved. We use a simple one-dimensional search over λ to find the minimum. Other approaches to the automatic choice of λ would be possible, for example the empirical Bayesian methods proposed by Leonard (1982).

5. Examples

Logistic Regression, and Tumour Prevalence Data

Dinse and Lagakos (1983) consider logistic regression models for data from a U.S. National Toxicology Program bioassay of a flame retardant. Data on 127 male and 192 female rats exposed to various doses of the agent consist of a binary response variable (y) indicating presence or absence of bile duct hyperplasia at death, and four explanatory variables: log dose (x_1), initial weight (x_2), cage position (x_3), and age at death (t). Dinse and Lagakos express some doubts as to whether the fourth of these variables enters the model linearly, so they consider fitting higher-order polynomials, or step functions based on age intervals. A reasonable alternative seemed to be the semi-parametric approach described here, which allows age at death (t) to enter the binomial logistic model in a non-parametric fashion, whilst still allowing estimation of the log dose regression coefficient.

The results of our analyses are presented graphically (see Figure 1) for various values of the tuning constant λ . In each plot, the upper two traces display on the same scale the fitted parametric and non-parametric components of the predictor, $\sum D_{ij}\hat{\beta}_j$ and $\hat{\gamma}(t) = \sum \hat{\xi}_k \phi_k(t)$, plotted against t_i . The lower panel displays the corresponding residuals from (8). (The slightly bizarre appearance of the residual plot is due to the binary nature of the response variable.) In both data sets there were a considerable number of ties in values of $\{ t_i \}$, which we have broken arbitrarily with small deterministic displacements. This seems not to lead to any numerical instability in our program, it avoided some modifications to the coding to handle coincident $\{ t_i \}$, and had the incidental advantage of clarifying the plots so that each case can be distinguished.

For the data on the male rats, Figures 1(a) and 1(b) demonstrate the effect of using very small and large values of λ respectively, suggesting under- and over-smoothing. These values are respectively 0.01 and 100 times the automatic GCV choice of $\lambda = 1380$, for which the relevant plot is Figure 1(c). This indicates a nonlinear dependence on t , but note that the left-hand part of the curve, up to the first turning value, is based on rather little data. The parameter estimates, with approximate standard errors, are -0.139 ± 0.148, -0.012 ± 0.022, and 0.068 ± 0.151, and we thus agree with Dinse and Lagakos about the lack of significance of the regression coefficients.

In the case of the female rats, the GCV method for choice of λ is not so well behaved. Whilst there is a turning value of Δ/v^2 at $\lambda = 6.24$, this is only a local minimum, and the GCV criterion seems to decrease to 0 for very small λ . Figure 1(d) displays the fitted values and residuals for $\lambda = 6.24$, suggestive of a complicated nonlinear dependence on t . Our parameter estimates are

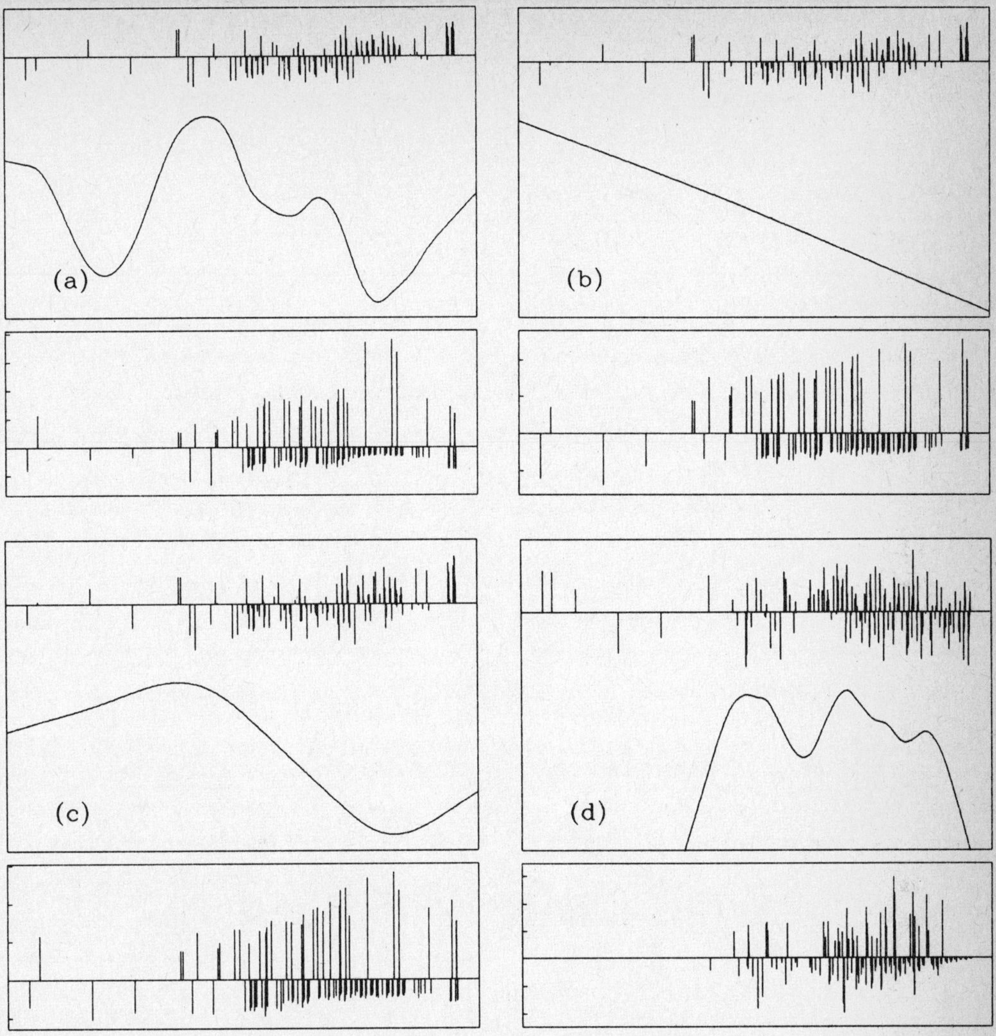

Figure 1. Fitted Logits and Residuals for Tumour Prevalence in Rats

0.492±0.137, 0.040±0.015 and 0.270±0.158. Dinse and Lagakos obtain $\hat{\beta}_1 = 0.554 \pm 0.20$ using a model linear in t, implying that if we have correctly identified the form of $\gamma(t)$ then their estimate is both slightly biased and inefficient.

It may be of interest to report some details of the performance of our algorithm applied to these data. For the male rats, with the GCV choice of λ, four iterations were needed to converge from initial estimates corresponding to empirical logits to a point where neither β nor the deviance changed by more than 10^{-4}. Excluding the calculation of $tr(\mathbf{S})$ (see section 4), the computation time was 1.26 seconds on a VAX 11/780. For the female rats, 10 iterations were required, and the time was 3.77 seconds.

Poisson Log-linear Regression, with Simulated Data

As a demonstration that our approach can properly identify a smooth $\gamma(t)$, we analyzed a simulated data set, with $\{ y_i \,;\, i = 1,2,...,200 \}$ distributed independently with Poisson distributions with means $\{ \exp(\theta_i) \}$, where

$$\theta_i = \sum_{j=1}^{5} D_{ij}\beta_j + 1 + \sin(t_i).$$

The $\{ t_i \}$ were chosen independently and uniformly on the interval $(0,3\pi)$, so that the true curve $\gamma(t) = 1 + \sin t$ has three turning values in the range of the data. The design matrix D was taken as that for 40 replicates of a randomized complete blocks design on 5 treatments, each successive 5 ordered $\{ t_i \}$ forming a block. The true β was (1, 0.5, 0, –0.5, –1), so that the Poisson means varied between about 0.4 and 20.

Figure 2(a) corresponds to the GCV choice of $\lambda = 3.13$: the fitted $\gamma(t)$ has the correct form, and the parameter estimates are (1.031, 0.517, 0.020, –0.583, –0.984) with standard errors (0.058, 0.066, 0.078, 0.098, 0.116). (Clearly for identifiability a constraint must be placed on β : we used $\sum\beta_j = 1$). Figure 2(b) demonstrates that when the tuning constant λ is set much too high, in this case 100 times the GCV value, the fitted curve $\gamma(t)$ cannot match the structure in the data, which is therefore forced into the residuals.

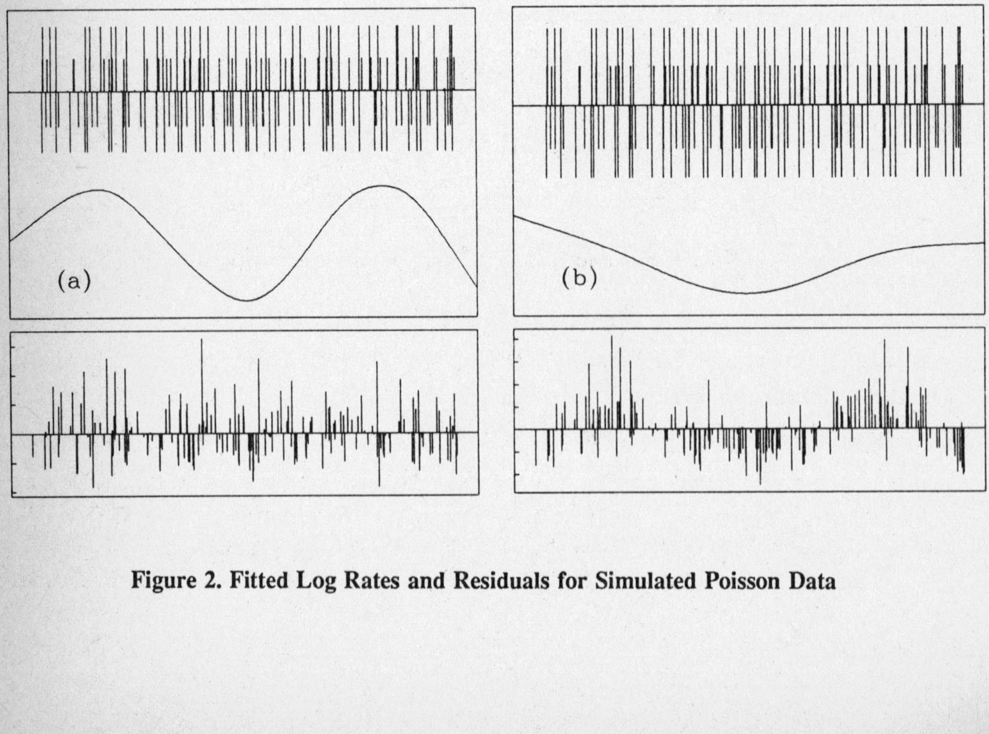

Figure 2. Fitted Log Rates and Residuals for Simulated Poisson Data

6. Using B-splines

For an application of spline smoothing to the penalized log-likelihood (4) more general than that of section 3, suppose that the $\{t_i\}$ remain one-dimensional, and that the roughness penalty takes the form

$$J(\gamma) = \int[\gamma^{(m)}(t)]^2 dt.$$

In this case, an alternative set of basis splines is useful, namely the natural B-splines of order $2m$ (De Boor, 1978; Schumaker, 1981). These are defined on any sequence of knots $s_1 < \cdots < s_q$ as piecewise polynomials of degree $(2m-1)$ between the knots, of degree $(m-1)$ outside (s_1, s_q), and with $(2m-2)$ continuous derivatives. This basis leads to stable, economical computing as B-splines are non-negative and have limited support. In the cubic spline case ($m = 2$), $\{\phi_k\,;\, k=3, \cdots, q-2\}$ are each non-negative only on (s_{k-2}, s_{k+2}), whilst ϕ_1, ϕ_2, ϕ_{q-1} and ϕ_q are linear outside (s_1, s_q). The matrix $\mathbf{K}$, which has the form

$$K_{ij} = \int \phi_i^{(m)}(t)\phi_j^{(m)}(t)dt,$$

is banded, and so an algorithm based on the equations (7) and (6) can be implemented in $O(n)$ time; see also Silverman (1985).

This approach is particularly useful in the case where n is very large and it is desirable that the number q of basis functions be much smaller. We are then only attempting the restricted minimization of (4), but with say $q = 50$ or 100 knots equally spaced to cover the range of $\{t_i\}$, this restriction is not of practical importance.

7. An Algorithm for the General Case

The linear system (5) can be expressed as finding $\boldsymbol{\beta}$ and $\boldsymbol{\xi}$ to minimize

$$\|\mathbf{B}^T(\mathbf{Y}-\mathbf{D}\boldsymbol{\beta}-\mathbf{E}\boldsymbol{\xi})\|^2 + \lambda\boldsymbol{\xi}^T\mathbf{K}\boldsymbol{\xi} \tag{11}$$

in which $\mathbf{A}=\mathbf{B}\mathbf{B}^T$. Suppose that $\mathbf{K}$ is of rank $r < q$. Two matrices, $\mathbf{J}$ and $\mathbf{T}$, with q rows and full column ranks r and $q-r$, respectively, can be formed such that $\mathbf{J}^T\mathbf{K}\mathbf{J} = \mathbf{I}$, $\mathbf{T}^T\mathbf{K}\mathbf{T} = 0$, and $\mathbf{J}^T\mathbf{T} = 0$ (see below). Rewriting $\boldsymbol{\xi}$ as

$$\boldsymbol{\xi} = \mathbf{T}\boldsymbol{\delta} + \mathbf{J}\boldsymbol{\varepsilon}, \tag{12}$$

with $\boldsymbol{\varepsilon}$ and $\boldsymbol{\delta}$ of lengths r and $q-r$, respectively, (11) becomes

$$\|\mathbf{B}^T(\mathbf{Y} - [\mathbf{D}{:}\mathbf{E}\mathbf{T}]\begin{pmatrix}\boldsymbol{\beta}\\ \boldsymbol{\delta}\end{pmatrix} - \mathbf{E}\mathbf{J}\boldsymbol{\varepsilon})\|^2 + \lambda\boldsymbol{\varepsilon}^T\boldsymbol{\varepsilon}.$$

A Householder decomposition (Dongarra et al., 1979) allows one to separate the solution of $\boldsymbol{\beta}$ and $\boldsymbol{\delta}$ from that of $\boldsymbol{\varepsilon}$. In other words, we decompose

$$\mathbf{Q}_1^T\mathbf{B}^T[\mathbf{D}{:}\mathbf{E}\mathbf{T}] = \mathbf{R},\ \ \mathbf{Q}_2^T\mathbf{B}^T[\mathbf{D}{:}\mathbf{E}\mathbf{T}] = 0,$$

in which $\mathbf{Q} = [\mathbf{Q}_1{:}\mathbf{Q}_2]$ is orthogonal and $\mathbf{R}$ is nonsingular, upper triangular, and of full rank $p+q-r$. The linear problem becomes: minimize the sum of

$$\|\mathbf{Q}_1^T\mathbf{B}^T\mathbf{Y} - \mathbf{R}\begin{bmatrix}\beta\\ \delta\end{bmatrix} - \mathbf{Q}_1^T\mathbf{B}^T\mathbf{E}\mathbf{J}\varepsilon\|^2 \tag{13}$$

and

$$\|\mathbf{Q}_2^T\mathbf{B}^T\mathbf{Y} - \mathbf{Q}_2^T\mathbf{B}^T\mathbf{E}\mathbf{J}\varepsilon\|^2 + \lambda\varepsilon^T\varepsilon. \tag{14}$$

The first term (13) can be set to zero by appropriate choice of β and δ given ε. If we define $\mathbf{Y}^* = \mathbf{Q}_2^T\mathbf{B}^T\mathbf{Y}$ and $\mathbf{Z} = \mathbf{Q}_2^T\mathbf{B}^T\mathbf{E}\mathbf{J}$, (14) becomes a problem of minimizing

$$\|\mathbf{Y}^* - \mathbf{Z}\varepsilon\|^2 + \lambda\varepsilon^T\varepsilon$$

which is an ordinary ridge regression problem. See Bates and Wahba (1983); Golub, Heath and Wahba (1979). The solution is

$$\varepsilon^* = (\mathbf{Z}^T\mathbf{Z}+\lambda\mathbf{I})^{-1}\mathbf{Z}^T\mathbf{Y}^*.$$

The other parameters are solved as

$$\begin{bmatrix}\beta^*\\ \delta^*\end{bmatrix} = \mathbf{R}^{-1}\mathbf{Q}_1^T\mathbf{B}^T(\mathbf{Y}-\mathbf{E}\mathbf{J}\varepsilon^*).$$

One then computes ξ^* using (12) and proceeds with the nonlinear iteration discussed in section 2.

Decomposition of K

One need only compute $\mathbf{J}$ and $\mathbf{T}$ once as $\mathbf{K}$ depends on the model only through $\{\mathbf{t}_i\}$. Using a pivoted Cholesky decomposition (Dongarra et al., 1979),

$$\mathbf{P}^T\mathbf{K}\mathbf{P} = \mathbf{L}^T\mathbf{L},$$

with $\mathbf{L}$ of dimension $r\times q$ and $\mathbf{P}$ a permutation matrix such that the first r columns of $\mathbf{KP}$ are linearly independent. A Householder decomposition of $\mathbf{L}^T$ yields

$$\mathbf{F}_1^T\mathbf{L}^T = \mathbf{G}, \;\; \mathbf{F}_2^T\mathbf{L}^T = 0,$$

in which $\mathbf{F} = [\mathbf{F}_1{:}\mathbf{F}_2]$ is orthogonal and $\mathbf{G}$ is nonsingular, upper triangular, and of full rank r. It is known (Dongarra et al., 1979) that $\mathbf{G}^{-1}\mathbf{F}_1^T$ is the Moore-Penrose inverse of $\mathbf{L}^T$. Therefore, one can construct the matrices $\mathbf{J}$ and $\mathbf{T}$ as

$$\mathbf{T} = \mathbf{P}\mathbf{F}_2 \text{ and } \mathbf{J} = \mathbf{P}\mathbf{F}_1\mathbf{G}^{-T}.$$

A further refinement is possible if the partial derivative matrices $\mathbf{E}$ and $\mathbf{D}$ can be written as $\mathbf{E} = \mathbf{E}^*\mathbf{M}$ and $\mathbf{D} = \mathbf{D}^*\mathbf{M}$, in which $\mathbf{M}$ depends only on $\{\mathbf{t}_i\}$. For instance, with a B-spline basis with model (1), $M_{ik} = \phi_k(t_i)$. One can left multiply by $\mathbf{M}$ exactly once, forming $\mathbf{T}^* = \mathbf{MT}$ and $\mathbf{J}^* = \mathbf{MJ}$ each with n rows, and replace $\mathbf{E}$ and $\mathbf{D}$ by $\mathbf{E}^*$ and $\mathbf{D}^*$ in subsequent computations.

Auxiliary Statistics

Auxiliary statistics can be constructed in the general case in a similar fashion to Section 4, with $\mathbf{S}$ in equations (9) and (10) replaced by

$$\mathbf{S} = \mathbf{E}(\mathbf{E}^T\mathbf{A}\mathbf{E} + \lambda\mathbf{K})^{-1}\mathbf{E}^T\mathbf{A}.$$

Alternatively, using the notation of this section, one can show that (9) becomes

$$\nu = n-p-q+r - tr\left[\mathbf{Z}^T\mathbf{Z}(\mathbf{Z}^T\mathbf{Z}+\lambda\mathbf{I})^{-1}\right].$$

The variance (10) can be reexpressed as

$$var\begin{bmatrix}\hat{\beta}\\ \hat{\delta}\end{bmatrix} = \mathbf{R}^{-1}\mathbf{Q}_1^T\left[\mathbf{I} + \mathbf{B}^T\mathbf{E}\mathbf{J}(\mathbf{Z}^T\mathbf{Z}+\lambda\mathbf{I})^{-1}\mathbf{Z}^T\mathbf{Z}(\mathbf{Z}^T\mathbf{Z}+\lambda\mathbf{I})^{-1}\mathbf{J}^T\mathbf{E}^T\mathbf{B}\right]\mathbf{Q}_1\mathbf{R}^{-T}.$$

O'Sullivan (1985) observed that the trace and the diagonal "leverage" elements of the hat matrix for the B-spline basis can be computed in $O(r)$ multiplications/divisions by using a Cholesky decomposition of ($\mathbf{Z}^T\mathbf{Z}+\lambda\mathbf{I}$) as in Silverman (1985).

8. Related Work in Progress

We have nearly completed an implementation of the algorithm for the general case, which will allow specification of models through subroutine evaluation of the forms of $\mathbf{A}$, $\mathbf{D}$, $\mathbf{E}$, $\mathbf{K}$ and $\mathbf{u}$. This software will be in the public domain and uses LINPACK (Dongarra et al., 1979) This work is related to a larger programming effort involving Douglas Bates, Grace Wahba, and Mary Lindstrom.

Yandell and Bates (unpublished) have also observed that one can use the singular value decomposition for ridge regression problems (Bates and Wahba, 1983; Golub, Heath and Wahba, 1979) to iterate on the "optimal" choice of tuning constant λ (in the sense of minimizing the generalized cross validation function) within each linear step. Thus λ changes with each nonlinear iteration, as do β and ξ . Empirically, convergence takes about the same number of steps as it would for a fixed λ near the optimal value. Unfortunately, the singular value decomposition is expensive, taking $O(r^3)$ multiplications/divisions. However, it may be possible to combine this approach with the Cholesky decomposition approach used by O'Sullivan, Yandell and Raynor, Jr. (1984) and Silverman (1985) to strike a healthy compromise between heavy computation and finding the optimal amount of smoothing.

Acknowledgements

This research has been supported in part by United States Department of Agriculture CSRS grant 511-100, National Sciences Foundation grant DMS-8404970, and United States Army contract DAAG29-80-C-0041. Computing was performed on the UW-Madison Statistics VAX 11/750 Research Computer and the UW-Madison MRC VAX 11/780 Computer.

References

Baker, R. J. and Nelder, J. A. (1978) *The GLIM System Release 3*. Oxford: Numerical Algorithms Group.

Bates, D. M. and Wahba, G. (1983) A truncated singular value decomposition and other methods for generalized cross-validation. Technical Report#715, Dept. of Statistics, U. of Wisconsin.

De Boor, C. (1978) *A Practical Guide to Splines*. New York: Springer.

Dinse, G. E. and Lagakos, S. W. (1983) Regression analysis of tumour prevalence data. *Appl. Statist.*, **32**, 236-248.

Dongarra, J. J., Bunch, J. R., Moler, C. B. and Stewart, G. W. (1979) *Linpack User's Guide*. Philadelphia: SIAM.

Eubank, R. L. (1984) The hat matrix for smoothing splines. *Statist. and Prob. Letters*, **2**, 9-14.

Eubank, R. L. (1985) Diagnostics for smoothing splines. *J. R. Statist. Soc. B*, **47**. (to appear)

Golub, G. H., Heath, M. and Wahba, G. (1979) Generalised cross validation as a method for choosing a good ridge parameter. *Technometrics*, **21**, 215-224.

Green, P. J., Jennison, C. and Seheult, A. H. (1983) Contribution to the discussion of the paper by Wilkinson et al. *J. R. Statist. Soc. B*, **45**, 193-195.

Green, P. J. (1984) Iteratively reweighted least squares for maximum likelihood estimation and some robust and resistant alternatives (with discussion). *J. R. Statist. Soc. B*, **46**, 149-192.

Green, P. J. (1985a) Linear models for field trials, smoothing, and cross-validation. *Biometrika*, **72**. (to appear)

Green, P. J. (1985b) Penalized likelihood for general semi-parametric regression models. Technical Report#2819, Math. Research Center, U. of Wisconsin.

Leonard, T. (1982) An empirical Bayesian approach to the smooth estimation of unknown functions. Technical Report#2339, Math. Research Center, U. of Wisconsin.

O'Sullivan, F., Yandell, B. S. and Raynor, Jr., W. J. (1984) Automatic smoothing of regression functions in generalized linear models. Technical Report#734, Dept. of Statistics, U. of Wisconsin.

O'Sullivan, F. (1985) Contribution to the discussion of the paper by Silverman. *J. R. Statist. Soc. B*, **47**. (to appear)

Reinsch, C. H. (1967) Smoothing by spline functions. *Numer. Math.*, **10**, 177-183.

Rice, J. R. (1981) An approach to peak area estimation. *J. Res. Nat. Bur. Stand.*, **87**, 53-65.

Schumaker, L. L. (1981) *Spline Functions: Basic Theory*. New York: Wiley.

Silverman, B. W. (1985) Some aspects of the spline smoothing approach to non-parametric regression curve fitting (with discussion). *J. R. Statist. Soc. B*, **47**. (to appear)

Wahba, G. (1977) A survey of some smoothing problems and the method of generalized cross-validation for solving them.. In *Applications of Statistics* (P. R. Krishnaiah, ed.), pp.507-523. Amsterdam: North Holland.

Wahba, G. (1984) Cross validated spline methods for the estimation of multivariate functions from data on functionals. In *Statistics: An Appraisal, Proceedings 50th Anniversary Conference Iowa State Statistical Laboratory* (H. A. David,, ed.) Iowa State U. Press.

Yandell, B. S. (1985) Graphical analysis of proportional Poisson rates. *Proceedings of the 17th Symposium on the Interface*, Lexington, Kentucky 17-19 March 1985.

AN ALGORITHM FOR DEGREE OF FREEDOM CALCULATIONS IN SPARSE COMPLETE CONTINGENCY TABLES

by Stephen J. Haslett
Institute of Statistics and Operations Research
Victoria University of Wellington, New Zealand

SUMMARY

The calculation of degrees of freedom for hierarchical loglinear models applied to complete contingency tables is usually straightforward. The calculation for such models applied to sparse complete multiway tables however, requires checking of the marginal totals for sampling zeros in all the configurations included in the model, together with an appropriate adjustment. GENSTAT, GLIM, SAS and SPSS^X do not make this adjustment. In this paper an algorithm is outlined which calculates degrees of freedom for loglinear models in such sparse tables. It has been shown that under suitable conditions, the distribution of the difference of the log likelihood ratio statistics for two nested loglinear models converges to the usual χ^2 distribution, and the algorithm then provides a routine method of calculating the appropriate degrees of freedom.

Keywords: Sparse; Complete; Multiway contingency tables; Hierarchical; Loglinear models; Degrees of freedom

1. INTRODUCTION

Categorical data problems have made use of chi-square (χ^2) goodness of fit statistics since 1900 with the pioneering work of Karl Pearson (Pearson, 1900). Review papers (e.g. Cochran, 1952) and papers on the asymptotic distribution of χ^2 statistics (e.g. Watson, 1959) appeared in the 1950's, and were followed by further developments in the analysis of categorical data including the use of loglinear models (e.g. Bishop et al, 1975, and Plackett, 1975, 1981).

The use of χ^2 goodness of fit statistics for large sparse multinomials is a more recent topic, and relevant central limit theorems have been developed by Morris (1975), Habermann (1977), and Koehler (1977). For loglinear models the usual goodness of fit statistic involves the likelihood ratio, and is

$$G^2 = 2N \Sigma \hat{p}_i \log \frac{\hat{p}_i}{\hat{\pi}_i} \qquad (1.1)$$

where the sum over the index i covers all elementary cells in the table, N is the total sample size, $\hat{p}_i$ and $\hat{\pi}_i$ are the maximum likelihood estimates (m.l.e's) under the saturated and test models respectively. G^2 is often called the deviance, as an abbreviation for the appropriate multiple of log likelihood ratio statistic given in equation (1.1). Williams (1976) has however shown that for large sparse tables G^2 is not asymptotically χ^2, but for complete multidimensional tables with closed m.l.e's. has derived a necessary scaling factor. Habermann (1977) has considered the related problem of comparison of two loglinear models for large sparse tables, and has shown that under appropriate conditions the difference of G^2 statistics between two nested loglinear models converges to the appropriate χ^2 distribution, despite neither G^2 value itself being distributed as χ^2. The degrees of freedom for the difference is the difference between the degrees of freedom for the two models. A review article on the use of χ^2 statistics for categorical data problems and including discussion of large sparse tables can be found in Fienberg (1979).

2. PRELIMINARIES

We begin with a few preliminary definitions which we will then relate to large sparse multiway tables, and more particularly to degree of freedom calculations.

A multidimensional contingency table is defined as complete iff, all elementary cells of the table may in principle contain non zero counts.

A multidimensional contingency table is defined as sparse iff it contains zero counts in at least one of its elementary cells.

Sparse complete p-dimensional tables with I_j levels to the jth factor, $j=1,2,..,p$, therefore contain at least one zero elementary cell count in the total of $\prod_{i=1}^{p} I_j$ elementary cells for which non-zero counts are possible. Such zero cells will be called sampling zeros and are to be distinguished from structural zeros which occur in cells of the table for which non-zero counts can never occur in any sample (e.g. grandmothers under the age of 5). Sampling zeros on the other hand are due to sampling variability, and disappear with increasing sample size.

The distinction betweeen sampling and structural zeros, while clear in principle and requiring different mathematical treatments, is not always clear in practice. For example, Newell, Ross and Renner (1984) collected more than fifty variables from a sample of New Zealand farms as a part of a study of small intestinal adenocarcinoma in sheep. Even if the fifty variables were dichotomous the number of cells in the corresponding contingency table would be 2^{50}, a far greater number than the number of sheep farms even in New Zealand; however given a particular sample of farms it is not clear which cells are sampling and which structural zeros. Such problems are not uncommon in sociological and epidemiological studies. In other cases however, for example, the study by Spagnuolo, Pasternack and Taranta (1971) on the recurrence of rheumatic fever, zero cell counts are assumed to be sampling zeros, a priori. The distinction between a structural and sampling zero is in fact one of existence (i.e. whether there is a non-zero probability of an observation in that cell); this can be a difficult distinction to draw without a priori information given only a finite sample.

We assume however in what follows that such a distinction can be made, and discuss only the situation where all zero cell counts are sampling zeros observed in a complete table; we further assume that all loglinear models for which degree of freedom calculations are to be made, are hierarchical.

Hierarchical loglinear models are discussed for example in Bishop, 1975 p. 34. Given a complete three-dimensional table IxJxK, the loglinear model

$$\log p_{ijk} = u + u_{1(i)} + u_{2(j)} + u_{3(k)} + u_{12(ij)}$$
$$i=1,2,..,I;\ j=1,2,..,J;\ k=1,2,..,K$$

is hierarchical, while the model

$$\log p_{ijk} = u + u_{2(j)} + u_{3(k)} + u_{12(ij)}$$

is not; here p_{ijk} is the probability associated with the (i,j,k)th cell of the table, $\{u, u_{1(i)}, u_{2(j)}, u_{3(k)}, u_{12(ij)}, u_{13(ik)}, u_{23(jk)}, u_{123(ijk)} : i=1..I; j=1..J; k=1..K\}$ is the set of all possible model effects and the unbracketed subscripts 1,2,3 denote the variables in the table. More generally for any multiway table, the family of hierarchical models is defined as the family such that if any u-term is set equal to zero, all its higher order relatives must also be set to zero. Thus if $u_1=0$ then $u_{12}=0$, and if $u_{12}\neq 0$ then $u_1\neq 0$.

Under the constraint that subscripted terms sum to zero across any associated index, the number of independent parameters for each u-term is less than the product of the associated categories, even for non-sparse tables, e.g. given a multiway table of dimension greater than two,

$$\sum_i u_{123(ijk)} = \sum_j u_{123(ijk)} = \sum_k u_{123(ijk)} = 0$$

and the number of independent parameters for the u-term, u_{123} in a non-sparse table is thus $(I-1)(J-1)(K-1)$, not IJK.

An alternative form of constraint is given for example in Plackett (1981), but is not discussed here in detail since it leads to the identical number of independent parameters for equivalent terms.

Generally for non-sparse p-dimensional tables the number of independent parameters for each u-term is found by subtracting one from the dimension of each component variable and finding their product, e.g. the variables denoted 1,2 and 3 are the component variables in the term u_{123} , and u_{123} has $(I-1)(J-1)(K-1)$ independent parameters for any p-dimensional table. Associated with each u-term, u_q is a corresponding configuration C_q where q denotes a sequence of indices corresponding to a subset of variables in the contingency table. Configurations are tables of sums obtained by summing over those variables in the table but not in the index set q. E.g. for the 3 dimensional table $C_{12} = \{x_{ij+} : i=1,2..I;\ j=1,2..J\}$ where $x_{ij+} = \sum_k x_{ijk}$ and x_{ijk} is the count in the (i,j,k)th cell. Configurations are sufficient statistics for their corresponding u-terms, under Poisson or simple multinomial sampling. The order, s, of a configuration C_q is the number of indices in the sequence q. Note that knowledge of a configuration C_q implies knowledge of all related configurations of lower order, since the sequence q' corresponding to any such configuration is made up only of elements of the sequence q. Such related configurations can be obtained from C_q by adding the elements of C_q over those indices in q but not in q'. Consequently given the u-terms contained in a particular loglinear model, a minimal sufficient set of configurations can be constructed.

3. DEGREE OF FREEDOM CALCULATIONS

Bishop et al (1975) give a procedure for calculating degrees of freedom for a loglinear model for a given non-sparse contingency table, namely

$$V = T_e - T_p \tag{3.1}$$

where V is the degrees of freedom associated with the particular model, T_e is the total number of cells in the table and T_p the total number of parameters fitted; T_p can be calculated by summing the number of independent parameters for each effect or u-term contained in the model. The method is extended in principle by Bishop et al (1975) to cover the case of sparse complete contingency tables for which

$$V = (T_e - Z_e) - (T_p - Z_p) \tag{3.2}$$

with V, T_e, T_p defined as before, Z_e equal to the number of elementary cells with zero estimates under the model, and Z_p the number of parameters that can not be estimated. Note that equations (3.1) and (3.2) are equivalent for non-sparse tables since for such tables $Z_e = Z_p = 0$.

The topic of this paper is determination of $(T_e - Z_e)$ the number of non-empty elementary cells under the model and $(T_p - Z_p)$ the total number of independent parameters in that model, for sparse complete tables. We note in passing that we do not discuss here the pathological situation where u-terms in the model rather than having zeros estimates cannot be estimated from the data, e.g. Bishop et al, 1975, Example 3.3-1, but note only that such invalid models are detectable since they give negative degrees of freedom when equation (3.2) is used.

We return to the non-pathological situation. Bishop et al (1975) make no comment on the general calculation of $(T_e - Z_e)$, but do make some preliminary comments on calculation of $(T_p - Z_p)$. They note that care must be taken in counting Z_p when more than one empty cell appears in a configuration, since if C_q of order s has adjacent zeros so arranged that the zeros persist in related configuration of order s-1, "then the zeros of C_q satisfy *some* of the constraints on the parameters of u_q ". Consequently if $Z(u_q)$ is defined as the loss of independent parameters from u_q then $Z(u_q)$ is less than the observed number of zeros, and the related u-terms of order s-1 also lose parameters when their related configurations contain zeros. By examining all configurations corresponding to u-terms in the model, they state that we may obtain Z_p as

$$Z_p = \sum_{q=1}^{N_u} Z(u_q)$$

where N_u is the number of subscripted u-terms in the model. Bishop et al (1975) do not however develop an explicit algorithm for calculations of Z_p and hence $(T_p - Z_p)$ for a loglinear model given any multiway sparse complete table.

Initially we discuss calculation of $(T_e - Z_e)$ the number of non-zero elementary cells under the model, since this is a matter rather quickly disposed of, provided iterative proportional fitting, IPF, (Deming and Stephan, 1940, Bishop et al, 1975, Section 3.5.2) is used to fit maximum likelihood estimates under the model.

Result 1:

For any loglinear model and multivariate table, the number of non-zero elementary cells estimates after one cycle of the IPF algorithm through the minimal set of configurations for that model is equal to $(T_e - Z_e)$, the number of non-zero elementary cell m.l.e's under the model.

Proof:

The general IPF procedure for fitting the minimal set of l configurations, C_{θ_r}, $r=1,2,..,l$ with cell entries x_{θ_r} respectively, is to fit initial elementary cell estimates $\hat{m}_\theta^{(0)}$ and proceed to fit each minimal configuration in turn, cycling until convergence of the elementary cell estimates. The usual procedure is to set $\hat{m}_\theta^{(0)} = 1$ for each elementary cell and to calculate $\hat{m}_{\theta_1}^{(0)}$ by appropriate summation. Generally after s cycles through the l minimal configurations,

$$\hat{m}_\theta^{(ls+1)} = \hat{m}_\theta^{(ls)}\, x_{\theta_1} / \hat{m}_{\theta_1}^{(ls)}$$

and at the tth step

$$\hat{m}_\theta^{(t)} = \hat{m}_\theta^{(t-1)}\, x_{\theta_r} / \hat{m}_{\theta_r}^{(t-1)}$$

where t-r is a multiple of l, so that during the first cycle, elements of $\hat{m}_\theta^{(r)}$ equal zero whenever the corresponding sums in x_{θ_r} equal zero. That is a zero in a minimal configuration results in zero estimates in the elementary cells which sum to give that zero configuration element, after only one cycle of the IPF algorithm. Further a zero elementary cell estimate cannot occur unless some corresponding marginal configuration contains a zero, since if an element of $\hat{m}_\theta^{(t-1)}$ is not equal to zero, the corresponding sum in x_{θ_r} must be zero for that element of $\hat{m}_\theta^{(t)}$ to become zero. Once set to zero at the $(t-1)^{th}$ step of the first cycle, elementary cell estimates contained in $\hat{m}_\theta^{(t-1)}$ remain zero in $\hat{m}_\theta^{(t)}$ and consequent steps, via the equation for the tth step.

$(T_e - Z_e)$ is thus easily calculated using the IPF algorithm as the number of non-zero cells after one complete cycle through the set of minimal configurations.

We now turn to calculation of $(T_p - Z_p)$, the number of independent parameters in the model. This we calculate as the sum of independent parameters associated with all the u-terms contained in the model, noting that each u-terms may be represented by its corresponding configuration.

Result 2:

(i) The number of independent parameters associated with a u-term, u_q, of order s, in a sparse multiway table is

$$T(u_q) = T(u_q)^* - Z(u_q) + \mathop{}_{q'} Z(u_{q'}) \tag{3.3}$$

where $T(u_q)^*$ is the number of independent parameters for u_q in the corresponding nonsparse table

$Z(u_q)^*$ is the number of zeros in the configuration C_q

$Z(u_{q'})$ is the number of independent parameters lost from the related configuration $C_{q'}$,

and related configurations to C_q, denoted $\{C_{q'}\}$, are formed from combinations of the elements or variables in q, taken 1,2,..,(s-1) at a time.

(ii) $$T_p - Z_p = \mathop{}_{q} T(u_q) \tag{3.4}$$

where the sum is over all configurations $\{C_q\}$ in the loglinear model.

Proof:

(i) Suppose that the number of independent parameters lost from each related configuration, $C_{q'}$, is known. Each independent parameter lost from some related configuration imposes a constraint on the zeros of C_q. Such lost parameters are mutually independent irrespective of the number of related configurations, $C_{q'}$, because such parameters are mutually independent in the associated nonsparse table. The zeros of C_q thus satisfy all the independent constraints imposed by the independent parameters. Letting $Z(u_q)^*$ denote the number of zeros in the configuration C_q we therefore have

$$Z(u_q) = Z(u_q)^* - \sum_{q'} Z(u_{q'}) \tag{3.5}$$

where the sum is over all the related configurations $C_{q'}$ of orders 0,1,2,..,s-1. Thus

$$T(u_q) = T(u_q)^* - Z(u_q) \tag{3.6}$$

is the total number of independent parameters for u_q for the sparse table. Combining equations (3.5) and (3.6) thus yields equation (3.3)

(ii) Summing $T(u_q)$ over all configurations, C_q, in the model thus yields

$$T_p - Z_p = \sum_{q} T(u_q)$$

We may thus calculate the number of independent parameters in any sparse complete multiway table for any given loglinear model using the following algorithm.

Algorithm:

1. Assuming the sparse table is not completely empty, determine the number of zero cells in each of the configurations of order 1 in the model. For such configurations denoted C_{q_1} the number of such zeros is $Z(u_{q_1})$. Set s=2.

2. Determine the number of zero cells in each configuration, C_{q_s} , of order s, in the model, and for each such configuration determine $Z(u_{q'})$ where the sum is over $\{C_{q'}\}$ the related configurations of order s-1,s-2,..,0 obtained by taking combinations of the elements of q_s. Hence determine $Z(u_{q_s})$, and $T(u_{q_s})$ via equations (3.5) and (3.6).

3. If s is the highest order of any configuration (i.e. u-term) in the model go to step 4 below, else increase the order s by 1 and return to step 2 above.

4. Calculate $(T_p - Z_p)$ via equation (3.4).

We consider two contrived examples.

Example 1:

Consider the 3x4x2 table with entries

$$C_{123} = \{3,5,7,10,2,17,1,12,0,0,0,0,0,0,0,0,0,0,0,0,0,0,0,0\}$$

$$= \{x_{111},x_{112},x_{121},x_{122},x_{131},x_{132},x_{141},x_{142},x_{211},.......,x_{342}\}$$

for which

$C_1 = \{57,0,0\}$ $\qquad C_2 = \{8,17,9,13\}$ $\qquad C_3 = \{13,44\}$

$C_{12} = \{8,17,19,13,0,0,0,0,0,0,0,0\}$ $\qquad C_{13} = \{13,44,0,0,0,0\}$

$C_{23} = \{3,5,7,10,2,17,1,12\}$

and

u_q	$T(u_q)^*$	$Z(u_q)^*$	$\sum_{q'} Z(u_{q'})$	$Z(u_q)$	$T(u_q)$
u	1	0	0	0	1
u_1	3-1=2	2	0	2	0
u_2	4-1=3	0	0	0	3
u_3	2-1=1	0	0	0	1
u_{12}	(3-1)(4-1)=6	8	2	6	0
u_{13}	(3-1)(2-1)=2	4	2	2	0
u_{23}	(4-1)(2-1)=3	0	0	0	3
u_{123}	(3-1)(4-1)(2-1)=6	16	2+6+2=10	6	0

Note that, not unexpectedly, $T(u_q) = 0$ for all u-terms, u_q , for which q contains variable 1. If we continue however to treat the contingency table as 3x4x2, sparse and complete, and fit various hierarchical loglinear models, designated by their minimal sufficient configurations, we find, for example the following degrees of freedom. Note that by way of illustration $\{C_1,C_{23}\}$ corresponds to the loglinear model, $\log p_{ijk} = u+u_1+(u_2+u_3+u_{23})$.

Model	$T_p - Z_p$	$T_e - Z_e$	df	G^2
$\{C_{12},C_{13},C_{23}\}$	1+0+3+1+0+0+3=8	24-16=8	8-8=0	0
$\{C_{13},C_{23}\}$	1+0+3+1+0+3=8	24-16=8	8-8=0	0
$\{C_1,C_{23}\}$	1+0+3+1+3=8	24-16=8	8-8=0	0
$\{C_1,C_2,C_3\}$	1+0+3+1=5	24-16=8	8-5=3	7.75

GENSTAT, GLIM, SAS and SPSS^x give the correct value for G^2 but give the degrees of freedom for these models as 6,12,14,17 respectively when treated as a 3-way table; we might thus easily come to the erroneous conclusion that $\{C_1,C_2,C_3\}$ is a satisfactory model for these data. For this example we may obtain the correct degrees of freedom using the standard computer algorithm for non-sparse complete tables by deleting the empty layers corresponding to the two values of the first variable for which there is no information, (i.e. conditioning on the first level of the first variable) and noting that the models corresponding to those above are then respectively, $\{C_{23}\}$, $\{C_{23}\}$, $\{C_{23}\}$ and $\{C_2,C_3\}$; however deletion of empty cells does not always give the correct degrees of freedom as the next example illustrates.

Example 2:

Consider the 3x4x2 table with entries

$$C_{123} = \{3,5,7,10,2,17,1,12,0,0,0,0,12,7,3,2,12,0,6,8,9,6,4\}$$

for which

$C_1 = \{57,24,53\}$ $\quad$ $C_2 = \{20,31,55,28\}$ $\quad$ $C_3 = \{60,74\}$

$C_{12} = \{8,17,19,13,0,0,19,5,12,14,17,10\}$ $\quad$ $C_{13} = \{13,44,15,9,32,21\}$

$C_{23} = \{15,5,13,18,22,33,10,18\}$

and

u_q	$T(u_q)^*$	$Z(u_q)^*$	$\sum_{q'} Z(u_{q'})$	$Z(u_q)$	$T(u_q)$
u	1	0	0	0	1
u_1	2	0	0	0	2
u_2	3	0	0	0	3
u_3	1	0	0	0	1
u_{12}	6	2	0	2	4
u_{13}	2	0	0	0	2
u_{23}	3	0	0	0	3
u_{123}	6	5	2	3	3

We consider to models $\{C_1,C_2,C_3\}$ and $\{C_{12},C_{13},C_{23}\}$ for which we calculate the correct degrees of freedom (d.f) for the complete sparse table, DF_S; the d.f for the table treated as non sparse, DF_{NS}; the d.f. supplied by GENSTAT and GLIM when the data are treated as 19 non-zero counts (i.e. zero cells are removed), DF_{NZ}.

Model	DF_S	DF_{NS}	DF_{NZ}	G^2
$\{C_1,C_2,C_3\}$	17	17	12	73.25
$\{C_{12},C_{13},C_{23}\}$	4	6	3	10.51

4. CONCLUSION

While not all sparse contingency table data should be treated as data from sparse complete tables, it is clear that standard computer packages for the analysis of categorical data using loglinear models (e.g. GLIM, GENSTAT, SAS, SPSS) do not calculate the correct degrees of freedom for such tables. Treatment of such sparse data as complete but non sparse results in overestimation of the degrees of freedom associated with a particular model, and hence underestimation of the mean deviance.

This is not however the only possible cause of mean deviances less than one, often observed for sparse tables, and "pseudo-sparse" tables, i.e. tables with relatively low but non-zero cell counts in some elementary cells. A further reason for such low mean deviances there, is that the number of zero or near-zero observed cells with estimates markedly different from their observed values decreases rapidly with increased model complexity particularly for sparse complete tables where zero fitted estimates for observed zero elementary cells then become common, and that for any model fit the major part of G^2 is contributed by elementary cells with low or zero observed counts but different fitted estimates. Consequently since maximum likelihood estimates for elementary cells with high observed counts are in such cases relatively insensitive to the model fitted, poor models fit very badly and good models very well.

Lin (1982, 1984) has developed computer programs in FORTRAN and in SAS for automatic loglinear model selection in complete, non-sparse contingency tables by a backward elimination procedure, having first established some satisfactorily overfitted model. The procedure, first discussed in Goodman (1971), is extendable to sparse, complete tables providing the necessary adjustments are made to the degree of freedom calculations for each model fitted, and a program written using the MATRIX procedure of SAS and which is a suitably modified form of that of Lin (1984), is available from the author.

REFERENCES

Bishop, Y.M.M., Fienberg, S.E., and Holland, P.W. (1975) *Discrete Multivariate Analysis*. Massachusetts: MIT Press.

Cochran, W.G. (1952) The χ^2 test of goodness of fit. *Ann. Math. Statist. 23*, 315-346.

Deming, W.E., and Stephan, F.F. (1940) On a least squares adjustment of a sampled frequency table when the expected marginal totals are known. *Ann. Math. Statist. 11*, 427-444.

Fienberg, S.E. (1979) The use of chi-squared tests for categorical data problems. *J. R. Statist. Soc. B. 41*, No. 1, 54-64.

Goodman, L.A. (1971) The analysis of multidimensional contingency tables: stepwise procedures and direct estimation methods for building models for multiple classifications. *Technometrics, 13*, No. 1, 33-61.

Habermann, S.J. (1977) Loglinear models and frequency tables with small expected cell counts. *Ann. Statist. 5*, No. 6, 1148-1169.

Koehler, K.J. (1977) Goodness of fit statistics for large sparse multinomials. Ph.D. dissertation, School of Statistics, University of Minnesota.

Lin, S.P. (1984) Automatic model selection in contingency tables. *SAS Communications X*, No. 2, 22-25.

(1982) Algorithm AS 185 Automatic model selection in contingency tables. *Applied Statistics 31*, 317-326.

Morris, C. (1975) Central limit theorems for multinomial sums. *Ann Statist. 3*, 165-188.

Newell, K.W., Ross, A.D., and Renner, R.M. (1984) Phenoxy and picolinic acid herbicides and small-intestinal adenocarcinoma in sheep. *Lancet*, December 8, 1301-1305.

Pearson, K. (1900) On a criterion that a given system of deviations from the probable in the case of a correlated system of variables is such that it can reasonably be supposed to have arisen from random sampling. *Phil. Mag. (5)*, 50, 157-75.

Plackett, R.L. (1974) *The Analysis of Categorical Data*. London: Griffin.

(1981) *The Analysis of Categorical Data*. 2nd Edn., London: Griffin.

Spagnuolo, M., Pasternack, B., and Taranta, A. (1971) Risk of rheumatic fever recurrences after streptococcal infections, prospective study of clinical and social factors. *New Eng. J. Med., 285*, 641-647.

Watson, G.S. (1959) Some recent results on chi-square goodness-of-fit tests. *Biometrics, 15*, 440-468.

Williams, D.A. (1976) Improved likelihood ratio tests for complete contingency tables. *Biometrika, 63*, 33-37.

GENERALIZED ADDITIVE MODELS; SOME APPLICATIONS

Trevor Hastie

Institute for Biostatistics

South African Medical Research Council

and

Robert Tibshirani

Department of Preventitive Medicine and Biostatistics

University of Toronto

Keywords and Phrases: GLIM, Smooth, non-parametric, logistic regression

SUMMARY

Generalized additive models have the the form $\eta(\mathbf{x})=\alpha+\Sigma f_j(x_j)$, where η might be the regression function in a multiple regression, or the logistic transformation of the posterior probability $p(y=1|\mathbf{x})$ in logistic regression. In fact, these models generalize the whole family of GLIM models $\eta(\mathbf{x})=\boldsymbol{\beta}'\mathbf{x}$ where $\eta(\mathbf{x})=g(\mu(\mathbf{x}))$ is some transformation of the regression function. We use the local scoring algorithm to estimate the functions, which uses a scatterplot smoother as a building block. The models are demonstrated in a non-parametric logistic regression. A variety of inferential tools have been developed to aid the analyst in assessing the relevance and significance of the estimated functions. The procedure can be used as a diagnostic tool for identifying parametric transformations of the covariates in a standard linear analysis.

1. INTRODUCTION

There are a number of advancements and enhancements for the linear regression model. The analyst is equipped with an ever growing toolbox of diagnostic checks for outliers and model deficiencies (see Belsley, Kuh and Welsch, 1980 and Cook and Weisberg, 1982).

Residual and partial residual plots are used to detect departures from linearity, and often suggest parametric fixes. An alluring alternative to this indirect approach is to model the regression function non-parametrically, and let the data decide on the functional form. For a single covariate the scatterplot smoother estimates the regression function in a local fashion (e.g. Reinch, 1967, Wahba and Wold, 1975, Cleveland, 1979). The parsimonious analyst will then let the smooth suggest a parametric form for the variable, and use the appropriate transformation in the linear regression model. Alternatively, one can make interpretations and predictions from the smooth itself.

A number of avenues have been opened for generalizing the scatterplot smooth to multivariate regression. An immdeiate generalization is to use a *surface* smoother. Friedman and Stuetzle (1981), amongst others, point out the dimensionality problems incurred when using surface smoothers. Essentially all smoothers base their estimates on some (weighted) average of the neighbouring observations. In high dimensions, one has to reach out further to find sufficient neighbours, and the estimate is no longer local. The same authors proposed the *projection pursuit regression* technique as an alternative to surface smoothing. This model has the form $E(Y|\mathbf{x}) = \Sigma_k f_k(\boldsymbol{\alpha}_k'\mathbf{x})$ where the f_k are estimated using a scatterplot smooth, and the $\boldsymbol{\alpha}_k$ are direction vectors onto which the data is projected. Projection pursuit techniques have received a lot of attention lately; for a recent overview see Huber (1985).

The *Additive Model*

$$E(Y|\mathbf{x}) = \alpha + \sum_{j=1}^{p} f_j(x_j), \qquad Ef_j(x_j)=0 \tag{1}$$

is a special case of the projection pursuit model in which the directions are preset at the coordinate directions. This model is less general than the projection pursuit model, but is more easily interpretable. As in linear regression, we can examine the effect of covariates one at a time, conditional on the presence of the other covariates. Only here we model the effects in a general non-parametric way. Alternatively, one can use the estimated functions in (1) to suggest parametric transformations for the covariates. A more traditional approach is to use residual and partial residual plots for this purpose; in this case non-linearities are detected in one of the covariates whilst all the others are kept linear. In (1) we can detect all the non-linearities simultaneously.

Pregibon (1981, 1982), and Landwehr, Pregibon and Shoemaker (1983) generalize the ideas in regression diagnostics to linear logistic regression problems, where the response variable y is zero or one. Partial residual plots are used to detect non-linearities in the model logit $p(\mathbf{x}) = \boldsymbol{\beta}'\mathbf{x}$. In this paper we describe and demonstrate the model

$$\text{logit } p(\mathbf{x}) = \alpha + \sum_{j=1}^{p} f_j(x_j), \qquad Ef_j(x_j)=0, \tag{2}$$

a generalized version of the additive model. With this model we can simultaneously detect non-linearities in all the covariates.

Nelder and Wedderburn (1972) and McCullagh and Nelder (1983) describe in detail the class of generalized linear models of the form $g(\mu(\mathbf{x}))=\mathbf{x}'\boldsymbol{\beta}$, where $\mu(\mathbf{x}) = E(y|\mathbf{x})$. This model includes the ordinary or euclidean regression model and the logistic regression model as special cases. The function $g(\cdot)$ is called the link function, and is usually assumed known. The model is estimated using an iteratively reweighted least squares procedure. If appropriate distributional assumptions are made, this is exactly maximum likelihood estimation; otherwise the procedure is justified on the basis of the quasi-likelihood.

Our generalization includes models of the form

$$g(\mu(\mathbf{x})) = \alpha + \sum_{j=1}^{p} f_j(x_j), \quad Ef_j(x_j)=0, \tag{3}$$

which we estimate using the *local scoring algorithm.* This is a natural generalization of the iterative least squares procedure for the linear problem. The methodology described in this paper applies directly to this general case. Further details can be found in Hastie and Tibshirani (1985a).

The local scoring algorithm is asymptotically equivalent to local likelihood estimation, another method for estimating models of the form (3) (Tibshirani, 1982, Hastie, 1983 and Tibshirani, 1984). Local Scoring has the advantage of being considerably faster. More recently O'Sullivan et al (1984) proposed a method of modelling generalized linear models in a non-parametric way using spline functions. This technique would yield estimates similar to ours in the one covariate case. To our knowledge, the additive model has not been used in this context; instead high dimensional splines are used, which are computationaly cumbersome and difficult to display and interpret.

To demonstrate the procedure, we perform a non-parametric logistic regression on some coronary risk factor data. This paper is largely contained in Hastie and Tibshirani(1985b), which has more detail, and includes a non-parametric analysis of covariance. The analysis was performed using GAIM, an interactive Fortran program (the name derives from GLIM (Baker and Nelder, 1978), with the obvious modification). A copy of the software is available on request from either author.

2. THE LOCAL SCORING ALGORITHM

In this section we simply give the algorithm, with motivations following in section 3. We use a bootstrap approach (as in booting up a computer). The most basic component, the *scatterplot smoother* , will be described first. Next we describe the *backfitting algorithm* which estimates the functions f_j in the model (1). Finally, the backfitting algorithm is used repeatedly at each iteration of the local scoring algorithm.

2.1 *The Scatterplot Smoother.*

The scatterplot smooth of a set of realizations (x_1,y_1), (x_2,y_2),..., (x_n,y_n) of a random pair (x,y) is a function $S(x_i|\mathbf{y},\mathbf{x})$ at each x_i, and is usually an estimate of $E(y|x=x_i)$. We will abreviate $S(x_i|\mathbf{y},\mathbf{x})$ to $S(y|x_i)$, $S(x_i)$ or even S when the context is unambiguous. A variety of smoothers exist in the literature. We use the local linear smoother (Cleveland, 1979, Friedman amd Stuetzle, 1982). These smoothers are intuitively motivated and computationally innexpensive (O(n) operations for the whole sequence).

$E(y|x=x_i)$ is the mean of y at x_i. In our sample we probably only have one observation at i, so instead we average all those y_j whose x_j lie in a neighbourhood of x_i. To overcome bias effects on the boundaries, the local linear smoother fits straight lines in the neighbourhoods, and picks $S(y|x_i)$ to be the value of the line at x_i :

$$S(x_i|\mathbf{y},\mathbf{x}) = a_i + b_i x_i \qquad (4)$$

where a_i and b_i are the estimated intercept and slope in the simple linear regression using the pairs (x_j,y_j), where $x_j \in N(i)=\{x_{i-k},x_{i-k-1},..., x_i,..., x_{i+k}\}$. Finally, we correct the smoother so that $\Sigma S(x_i|\mathbf{y},\mathbf{x})=\Sigma y_i$.

The number k is a parameter of the smoother, and determines the size of the neighbourhood. We refer to the number (2k+1)/n as the span. Large spans produce smooth curves high in bias and low in variance, whereas small spans tend to produce wiggly curves low in bias and high in variance. We can pick the span by cross-validation (Friedman and Stuetzle, 1982), Stone (1974). In practice we tend to use arbitrary spans for data exploration, a typical value of k being 2k+1 = n/2 or a span of 50%. The chief advantage of this smoother is speed; notice that the defining formulae for (4) are easily updatable from one neighbourhood to the next, since only two points change. This feature makes the local linear smoother a popular choice as a primitive in more complex algorithms such as ours, where repeated smooths are required. We emphasise, however, that in what follows the operator $S(y|x)$ can be replaced by any regression type scatterplot smoother. Finally, the notation $S(y,w|x)$ will refer to a *weighted* smooth. In this case, there is a weight w_i associated with each observation; all the averages in (4) become weighted averages.

2.2 *The Backfitting Algorithm*

The backfitting algorithm estimates the functions f_j in the model $E(y|\mathbf{x}) = \alpha+\Sigma f_j(x_j)$.

<u>Backfitting Algorithm</u>

<u>Initialization</u> $f_j \equiv 0$, $\alpha = ave(y_i)$.

<u>Cycle</u> j=1,2,...,p,1,2,...,p, 1,2,...

$$r_i = y_i - \alpha - \sum_{k \neq j} f_k(x_{ki}) \quad i=1,..n$$

$$f_j(x_{ji}) = S(r \mid x_{ji}) \quad i=1,..n$$

Until $$RSS = \sum_{i=1}^{n} (y_i - \alpha - \sum_{j=1}^{p} f_j(x_{ji}))^2 \text{ converges.}$$

At each stage the algorithm smooths residuals against the next covariate; these residuals are obtained by removing the estimated functions or covariate effects of all the other variables.

The backfitting algorithm has appeared a number of times in the literature. Andrews et al (1967) used a similar algorithm (Multiple Classification Analysis) in fitting additive effects in an Anova situation. Friedman and Stuetzle (1981) refer to this special case of projection pursuit regression as "projection selection". The algorithm has the same flavour as the iterative proportional scaling algorithm used in fitting multiplicative models to multiway contingency tables (eg see Fienberg, 1981, Hastie and Tibshirani, 1985c). Stone(1984) gives a thorough historical account of the additive model and the algorithm, and proves some results on rates of convergence.

Some of the properties of the algorithm are:

° If the S(.) refer to least squares fits, the algorithm converges to the least squares solution (Hastie and Tibshirani, 1985a).

° Breiman and Friedman (1982) prove a consistency result for the algorithm. They also show that for a class of simple but unpractical smoothers, the algorithm converges. In practice, the algorithm always has converged.

° One can view the algorithm from the viewpoint of the model, where we replace $S(y \mid x)$ by $E(y \mid x)$. In this case, under mild regularity assumptions:

° An unique additive approximation to $E(y \mid \mathbf{x})$ of the form $\alpha + \Sigma f_j(x_j)$ exists, and the algorithm converges to it (Breiman and Friedman, 1982, Stone, 1984).

° Under further regulatory assumptions on the joint distribution of the x_j, the individual functions f_j are unique.

2.3 *The Local Scoring Algorithm*

We give the algorithm for the logistic regression problem (2); Hastie and Tibshirani (1985a) present the algorithm in full generality. We have a sample $(y_1, \mathbf{x}_1), \ldots, (y_n, \mathbf{x}_n)$, where $y_i = 0$ or 1 and $\mathbf{x}_i$ is a vector of p covariates.

<u>The Local Scoring Algorithm</u>

<u>Initialization</u> $f_j^0 \equiv 0$, $j=1,..p$, $\alpha = \text{logit}(\text{ave}(y))$.

<u>Loop</u> over outer iteration counter m

$$\eta^m(\mathbf{x}_i) = \alpha + \sum_{j=1}^{p} f_j^m(x_{ji})$$

$$p_i = \text{logit}^{-1}(\eta^m(\mathbf{x}_i))$$

$$= \exp(\eta^m(\mathbf{x}_i))/[1 + \exp(\eta^m(\mathbf{x}_i))]$$

$$z_i = \eta^m(\mathbf{x}_i) + (y_i - p_i)/[p_i(1-p_i)], \quad w_i = p_i(1-p_i)$$

Obtain $f_j^{(m+1)}$, $j=1,..p$ by applying the backfitting algorithm to z_i, with observation weights w_i and starting values f^m_j.

<u>until</u> the deviance $D(\mathbf{y},\mathbf{p}) = -2\Sigma\,[y_i \ln p_i + (1-y_i) \ln(1-p_i)]$ converges.

We make the following observations:

° If the backfitting algorithm is replaced by the overall weighted least squares fit, this algorithm is identical to the iteratively reweighted least squares (IRLS) algorithm for solving the maximum likelihood equations in linear logistic regression (see McCullagh and Nelder, 1983, for an account of the IRLS algorithm). If our smoother fits global (weighted) linear fits, then from the previous result we see that the algorithm convergesto the MLE for the linear logistic model.

° A naive approach to this problem might be to model logit (y_i) by an additive model, possibly with weights. Since y_i is 0-1, we cannot take logits; instead the algorithm computes a first order Taylors series approximation z_i to logit (y_i) about the point p_i, the best current guess to $E(y_i|\mathbf{x}_i)$. The z_i are then modelled by an additive model, and the process is repeated. Indeed, if $\eta^0(\mathbf{x})$ is the current guess for the true η, the local scoring procedure estimates a solution to the weighted least squares problem :

$$\text{MIN } Ew^0(\mathbf{x})[\eta^0(\mathbf{x})+(y-p^0(\mathbf{x}))/p^0(\mathbf{x})(1-p^0(\mathbf{x}))-\alpha-\Sigma f_j(x_j)]^2 \qquad (5)$$

where $w^0=p^0(1-p^0)$.

° Landwehr et al (1983) proposed smoothing partial residuals from the linear fit to detect non-linearities. The partial residual for variable j and observation i is $r_{ij}= b_j x_{ij}+(y_i-p^l_i)/p^l_i(1-p^l_i)$, where p^l_i is the fit from the linear model. This is exactly the first step of the backfitting procedure within the local scoring algorithm, if we start with the linear fit. The local scoring procedure goes on to estimate all the non-linearities simultaneously (see also Hastie, 1983).

The local scoring algorithm produced the functions in figures 4.1-4.3. All smooths used a span of 50%. The next section gives some details of the algorithm, as well as a theoretical motivation. In the first pass, the reader might skip to section 4 where we use the procedures on some real data.

3. DETAILS OF THE LOCAL SCORING ALGORITHM

We first motivate the algorithm for the case of a single covariate. The scatterplot smooth estimates $E(y|x)$, the theoretical minimizer of the mean squared prediction error

$$EPE = E[y-f(x)]^2 = E(y-E(y|x))^2+E(f(x)-E(y|x))^2. \qquad (6)$$

Often we write EPE conditioned on the sample values of x, in which case it is $EPE=\Sigma_i E[(y-f(x_i))^2|x_i]$. EPE estimates the ability of f(x) in predicting future observations y at x, averaged over x. The use of the cross-validated residual sum of squares (RSS) for span selection emphasises that our goal is, in fact, to minimize EPE (Stone, 1974).
Generalized linear models are usually estimated by maximum likelihood within the exponential family. If we write the log likelihood of the modelled mean $\mu(x_i)$ as $l(y_i,\mu(x_i))$, then the parameters in the linear model $g(\mu(x))=\eta(x)=\alpha+\beta x$ are chosen to minimize $L=\Sigma_i\, l(y_i,\mu(x_i))$. We can now generalize the ideas above and seek an estimate $\mu(x)$ to minimize $ELL=El(y,\mu(x))$, the expected log-likelihood of future observations. Alternatively, let $K(\mu_1,\mu_2)$ denote the Kullback-Leibler distance (KL) between two exponential family densities, one with mean μ_1 and the other with mean μ_2:

$$K(\mu_1,\mu_2) = E_{\mu_1} \log (f_{\mu_1}(x)/f_{\mu_2}(x))$$

where f_{μ_1} and f_{μ_2} are the two densities.

We can think of $K(y,\mu)$ as measuring the loss in using μ to predict y. (We will interchangeably write $K(y,\mu)$ or $K(y,\eta)$, where $\eta=g(\mu)$ is the natural parameter corresponding to μ). Hoeffding's lemma (Efron, 1977) in fact shows us that $K(y,u)=l(y,y)-l(y,u)$. So maximizing ELL corresponds to minimizing

$$\begin{aligned} EKL &= EK(y,\mu(x)) \\ &= EK(y,\mu^0(x)) + EK(\mu^0(x),\mu(x)), \end{aligned} \tag{7}$$

where $\mu^0(x)=E(y\,|\,x)$, the true regression function. It is clear that, once again, μ^0 solves the problem. If $\mu(x)$ is a constrained family, then we find the member of the family closest in EKL to $\mu^0(x)$. Either way (ELL or EKL), a sufficient condition for η to minimize EL is given by $E(dl/d\eta\,|\,x)_{\eta(\mathbf{x})}=0$. Given a guess $\eta_1(x)$, an improved guess is

$$\eta_2(x) = E(\eta_1(x) - [dl/d\eta_1]/E[d^2l/d\eta_1^2\ |x]\ |\,x) \tag{8}$$

using standard Newton-Raphson arguments. For the generalized models this simplifies to $\eta_2(x)= E(\eta_1(x) + (y-\mu_1(x))\,[d\eta/d\mu]_1\ |\,x)$, and for the logistic regression model to $\eta_2(x)= E(\eta_1(x)+(y-p_1(x))/p_1(x)(1-p_1(x))\ |\,x)$. The data algorithm simply replaces the conditional expectation by an estimate, the scatterplot smooth.

For multiple covariates, $\eta(\mathbf{x})=\alpha+\Sigma_j f_j(x_j)$, so the intuitive modification is to replace $E(\cdot\,|\,\mathbf{x})$ by the backfitting algorithm, estimated using smooths. From a theoretical viewpoint, a little algebra shows that convergence of the local scoring algorithm (at the model) implies that

$$E[\delta l(y,\eta_1(\mathbf{x}))/\delta f_{1j}(x_j)\ |\,x_j) \equiv 0\ \ \forall\, j. \tag{9}$$

This is a sufficient condition for $\eta_1(\mathbf{x}) = \alpha+\Sigma_j f_{1j}(x_j)$ to be the additive model closest in EKL or EL to $\eta^0(\mathbf{x})$. (See Stone, 1985, for results on the existence and uniqueness of such additive approximations in the exponential family). For the logistic model, this condition is $E(y-p_1(\mathbf{x})\,|\,x_j)\equiv 0\ \forall\, j$, which has the intuitive interpretation that model and true marginals must agree along all co-ordinate directions. These are the theoretical analogues of the usual score equations in maximum likelihood estimation.

5. NON-PARAMETRIC LOGISTIC REGRESSION

The data in this example is a subset of the Coronary Risk Factor Study (CORIS) baseline survey, carried out in three rural areas of the Western Cape, South Africa. This part of the study aims to identify and establish the intensity of ischaemic heart disease (IHD) risk factors in this high incidence region. The baseline study is to be followed by a two level intervention program (Rossouw et al, 1983). We analyze the data for the 3357 white males between tha ages 15 and 64, and concentrate on risk factors for mycardial infarction (MI). The overall incidence of MI in 1979 was 5.13% for this group.

Denote the presence or absence of MI for observation i by the binary variable y_i. Initially a large number of possible risk factors and functions thereof were considered; these were later reduced to a set of possibly significant factors using stepwise linear logistic regression techniques (We thank D. Capatos for making his analysis available to us, and J. Rossouw for allowing us to use these data). We used this set as the starting point in our analysis. The set of risk factors are:

1. Systolic Blood Pressure

2. Cumulative Tobacco(in kg): attempts to measure the total tobacco consumed in the subject's lifetime, and is simply the average per day multiplied by the period of use.

3. Cholesterol Ratio : defined as (Total-HDL)/HDL where Total is the overall serum cholesterol measurement, and HDL is the amount of High Density Lipoprotein. It is fairly well accepted that the "good" HDL tends to counteract the "bad" LDL, and so this ratio becomes relevant.

4. Type A : is a measure of phsycho-social stress, as measured by the self administered Bortner Scale.

5. Age

6. Total Energy: is a measure of the total energy expended in leisure time and occupational activities.

7. Family History: is a zero-one variable, where a one indicates that a family member of the subject has had heart disease.

In our analysis we used the 162 cases, and sampled roughly double the number (303) from the "controls". As pointed out by Anderson(1972), such sampling still allows us to model the contribution of each risk factor to the log-odds of risk without bias; the overall level of risk (the constant term) is, however, distorted.

We used the local scoring procedure to estimate the model

$$\text{Logit } p(\mathbf{x}) = \alpha + \sum f_j(x_j)$$

where x_j , j=1,..7 denote the seven covariates. A span of 0.5 was used for all the covariates except family history, where we used a span of 1. This amounts to linear regression for this variable, which in this case estimates a contrast. Some of the fitted functions are given in figures 4.1 through 4.3, together with +-2SD curves. The analysis of deviance (ANODEV) table 4.1 summarizes the contribution of each variable to the fitted model. Each entry in the table corresponds to the increase in the deviance as a result of the exclusion of that term from the full model. Details of the computation and justification for the SD (standard deviation) curves and the *degrees of freedom* or equivalent number of parameters is given in Hastie and Tibshirani(1985a, b). The plots also contain the maximum likelihood estimates for the linear terms in the model logit $P(\mathbf{x}) = \beta_0 + \mathbf{x}'\boldsymbol{\beta}$ for comparison. These are represented by dashed lines. The deviance for this model is 463.4 with 8 parameters or degrees of freedom (dof) in the estimate. The estimated additive model has a deviance of 440.7 for 17.7 dof, a drop of 22.7 for 9.7 dof. The figures also include step functions. Each continuous variable was divided into a suitable number of categories. In our case we chose the quintiles; the variable cumulative tobacco was broken into 3 categories since 40% of the observations recorded a value of 0. A separate constant is estimated for each category in the model

$$\text{Logit } P(\mathbf{x}) = \beta + \beta_{11}I_{11}(x_1) + \beta_{12}I_{12}(x_1) + .. \beta_{15}I_{15}(x_1) + \beta_{21}I_{21}(x_2) + ... \beta_{75}I_{75}(x_7)$$

where the I_{jk} is an indicator variable for category k of variable j. The constants model has a deviance of 472.9 for 24 dof, a worse fit than the linear model with 3 times the number of parameters! This is not too surprising, since a number of the non-paramteric fits were fairly linear (the ones not included here), and the constants model will pick up bias in these cases. We note, in addition, that it is not as easy to spot non-linearities using step functions as it is with the non-parametric curves. It is also clear that category choice can play an important role in the constants model. A row of stars is plotted at the base of each curve; these represent the occurrence of data points, although the frequency is not represented. We note that the width of the confidence curves at the upper range in figures 4.1 and 4.2, for example, will be due mainly to the sparcity and outlying nature of the observations in that region.

We now provide some interpretations and further analysis of the various effects:

<u>Systolic Blood Pressure</u>

Exclusion of the linear term for systolic blood pressure (SBP) from the linear model causes the deviance to increase by 1.8 to 465.2. The deviance for the additive model, on the other hand, increases by 11.4 for 2.8 dof when SBP is excluded. Thus the linear model is unable to detect a significant effect for SBP. This is not surprising, since in figure 4.1 we see that the non-parametric estimate is U shaped. Figure 4.4a is a partial residual plot for SBP. Our partial residuals are a natural extension of those of Landwehr et al (1983) as defined in section 2: $r_{ij} = f_j(x_{ij}) + (y_i - p(\mathbf{x}_i))/[p(\mathbf{x}_i)(1-p(\mathbf{x}_i))]$. It was discovered that 91 of the 465 people in this subset of the data were on treatment for high blood pressure. A possible explanation for the U shape curve is that people who have a MI subsequently go on treatment for high blood pressure, and their pressure drops. We have coded the treatment variable into the partial residual plots (+ is treatment, 0 no treatment), and note that the +'s tend to be above the curve. More conclusive evidence is provided in figure 4.4b where we fit a separate SBP curve for the two groups. As expected, the no treatment SBP curve is increasing; the treatment curve adds the contamination which results in the U shape. The fit is shown on the probability scale (as apposed to logit scale), since for this illustration we have not adjusted for the other covariates. This also demonstrates the ability of the local scoring procedure in uncovering the regression in a scatterplot with little visual information. One would naturally include separate functions for the two groups in any further analysis of the joint behaviour of the covariates.

Cholesterol Ratio

As in the above, we can code any information in the partial residual plots. Figure 5.8a is a plot for Cholesterol Ratio (ChR), with the variable Family History (+ = yes) encoded. There is a predominance of 0's above the curve, and +'s below. Figure 4.8b shows a marginal fit for the two groups, and the interaction is clearly demonstrated. The original model attempted to capture the different sloped curves by a single curve with two intercepts.

Cumulative Tobacco

This curve (figure 4.2) shows a sharp increase in the logit of risk from people who have never smoked to people who have smoked at all; from then on the risk increases gradually with amount smoked.

6. DISCUSSION

Generalized additive models provide a flexible method for identifying non-linear covariate effects in a variety of modelling situations; notably the very situations in which it has become popular to use the generalized linear or GLIM models. The additive models can be used in a data analytic fashion to understand the effect of covariates, and to test hypothesis about effects. Alternatively, one can allow the non-parametric functions suggest parametric transformations, and then proceed with the usual linear analysis on the transformed variables.

Some theory already exists for these models, including results on existence, uniqueness, and convergenceat the model, and consistency. We can estimate the degrees of freedom of the terms in the model and calculate approximate standard deviation curves for the fitted functions. This theory is still developing, with a variety of relevant questions remaining to be answered. One such example is the effect of dependence of the covariates on the fitted algorithm, the standard deviations, and the degrees of freedom of the fit. Partial progress has been made in this direction.

We have illustrated the procedures on binary data here; Hastie and Tibshirani(1985b) perform a non-parametric analysis of covariance. Hastie and Tibshirani (1985a) analyzed survival data, and used an extension of the ideas to estimate in addition a non-parametric link function. Obvious extensions are under development to include two dimensional surface interaction terms in the models. The algorithm is straightforward, and can be incorporated into standard packages such as SAS, S and GLIM.

REFERENCES

Anderson, J. (1972) Separate Sample Logistic Regression, *Biometrika*, **59**, 19-35.

Andrews, F., Morgan, J., and Sonquist, J. (1967) Multiple Classification Analysis, *Inst . Social Research*, University of Michigan, Ann Arbor.

Baker, R. and Nelder, J. (1978) *The GLIM System, release 3*, Distributed by NAG, Oxford.

Belsley, D., Kuh, E., and Welsch, R. (1980) *Regression Diagnostics*, Wiley, New York.
Breiman, L. and Friedman, J. (1982) Estimating Optimal Transformations for Multiple Regression and Correlation, *Dept. Stat. tech. rep., Orion 16*, Stanford University.
Cleveland, W. (1979) Robust Locally Weighted Regression and Smoothing of Scatterplots, *J. Amer. Statist. Assoc.*, **74**, 829-836.
Cook, R. and Weisberg, S. (1982) *Residuals and Influence in Regression*, Chapman and Hall, London.
Efron, B. (1977) The Efficiency of Cox's likelihood function for censored data. *J. Amer. Statist. Assoc.* **72**, 557-565.
Fienberg, S. (1981) *The Analysis of Cross-Classified Categorical Data*, MIT press.
Friedman, J. and Stuetzle, W. (1981) Projection Pursuit Regression, *J. Amer. Statist. Assoc.*, **76**, 817-823.
Friedman, J. and Stuetzle, W. (1982) Smoothing of Scatterplots, *Statistics Dept. Tech. rep.*, Orion 3, Stanford University.
Hastie, T. (1983) Non-Parametric Logistic Regression, *Statistics Dept. Tech. rep.*, Orion 16, Stanford University.
Hastie, T. and Tibshirani, R (1985a) Generalized Additive Models. (to appear in *J. Statist. Science.*)
Hastie, T. and Tibshirani, R. (1985b) Generalized Additive Models; some Applications, submitted for publication.
Hastie, T. and Tibshirani, R (1985c) Comment in "Huber (1985)".
Huber, P. (1985) Projection Pursuit, *Annals Statist. (to appear).*
Landwehr, J., Pregibon, D., and Shoemaker, A. (1982) Graphical Methods for Assessing Logistic Regression Models, *J. Amer. Statist. Assoc.*, **79**, 61-63.
McCullagh, P. and Nelder, J. (1983) Generalized Linear Models. Chapman Hall, London.
Nelder, J. and Wedderburn, R. (1972) Generalized Linear Models, *J. R. Statist. Soc. A*, **135**, 370-384.
O'Sullivan, F. Yandell, B., and Raynor, W. (1984) Automatic Smoothing of Regression Functions in Generalized Linear Models, *Dept. Stat. tech. rep. 734*, University of Wisconsin
Pregibon, D. (1981) Logistic Regression Diagnostics, *Annals of Statistics*, **9**, 705-724.
Pregibon, D. (1982) Resistent Fits for Some Commonly Used Logistic Models with Medical Applications, *Biometrics*, **38**, 485-498.
Reinsch, C. (1967) Smoothing by Spline Functions, *Numer. Math.*, **10**, 177-183.
Rossouw, J., du Plessis, J., Benade, A., Jordaan, P., Kotze, J., Jooste, P., and Ferreira, J. (1983) Coronary Risk Factor Screening in Three Rural Communities, *South African Med, J.*, **64**, 430-436.
Stone, C. (1984) Additive Regression and other Non-Parametric Models, *Statist. dept. tech. rep. 33*, U. of Berkely, California.
Stone, C. (1985) Personal Communication.
Stone, M. (1974) Cross-validatory Choice and Assessment of Statistical Predictions (with Discussion). J. *R. Statist. Soc. B*, **36**, 111-147.
Tibshirani, R. (1982) Non-Parametric Estimation of Relative Risk, *Statistics Dept. Tech. rep.*, Orion 22, Stanford University.
Tibshirani, R. (1984) Local Likelihood Estimation. Unpublished Ph.D thesis, Stanford University.
Wahba, G., and Wold, S. (1975) A Completely Automatic French Curve: Fitting Spline Functions by Cross-Validation, *Comm. Statistics*, **4**, 1-7.

Figure 4.1

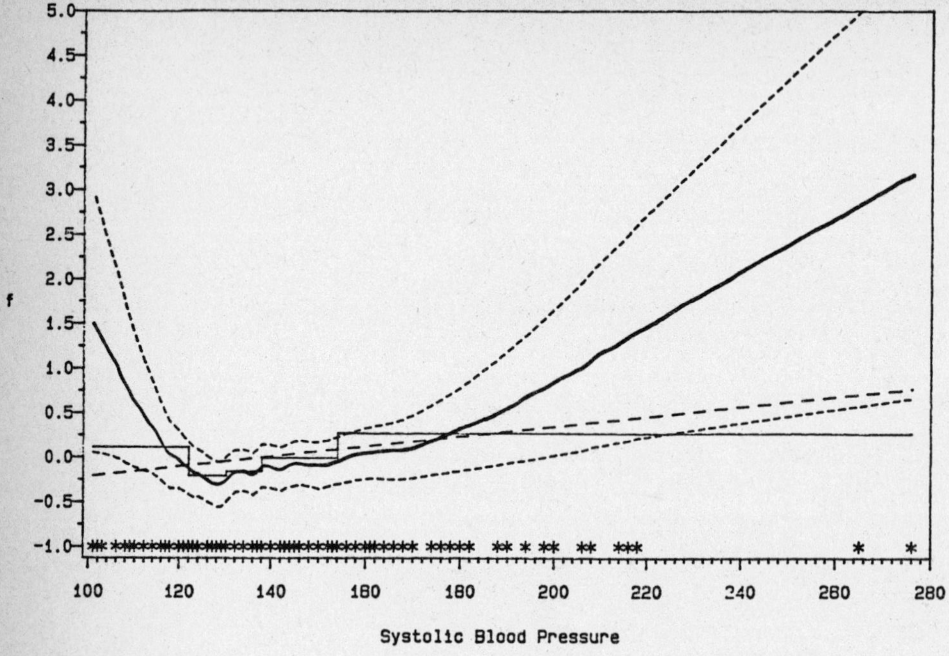

Figure 4.2

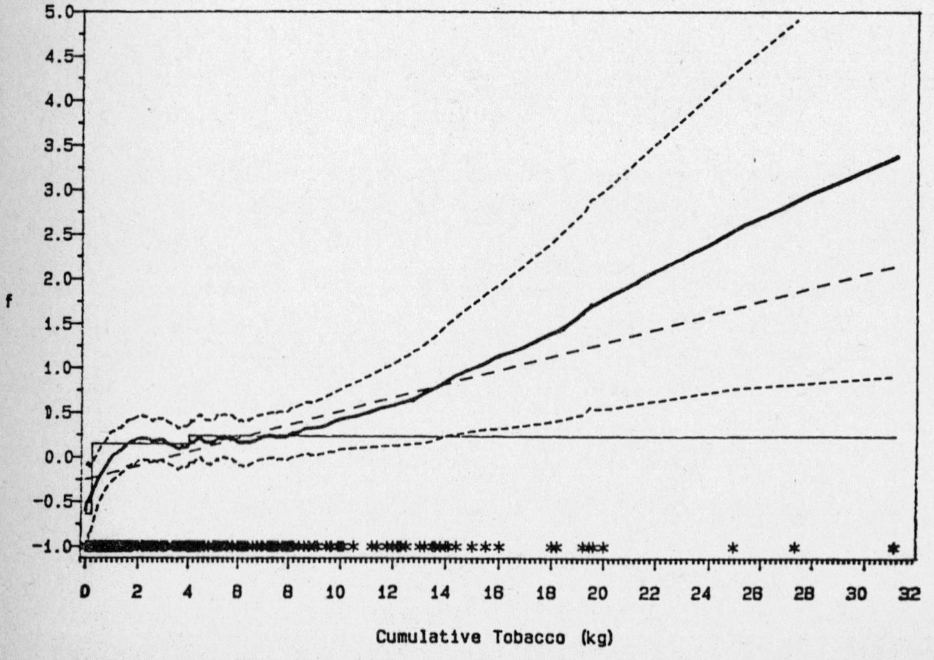

Figure 4.3

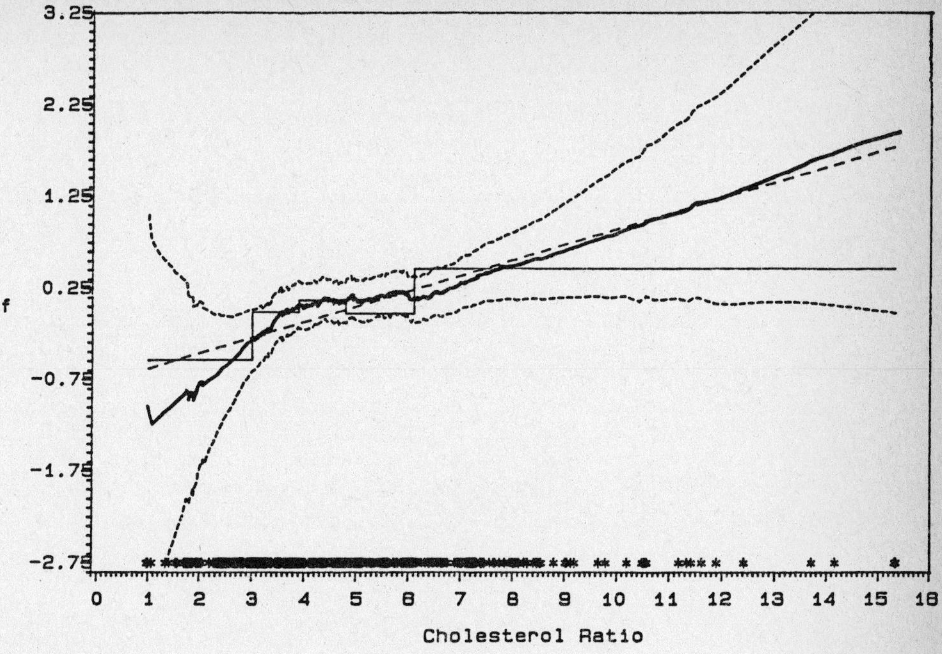

Table 4.1 - ANODEV

TERM	DEVIANCE	DOF	EFFECT	DOF
Linear in 7 covariates	463.4	8		
above less SBP	465.2	7	1.8	1
Constants model	472.9	24		
Full additive model	440.7	17.7		
full les SBP	452.3	14.9	11.4	2.8
full less cum. tobacco	448.2	15.2	7.5	2.5
full less cholesterol R	453.0	14.9	12.3	2.8
full less Type A	460.1	15.0	19.4	2.7
full less age	455.4	15.2	14.7	2.5
full less total energy	456.9	15.1	16.2	2.6
full less family history	460.5	16.7	19.8	1

Figure 4.4a

Partial residuals and Treatment variable

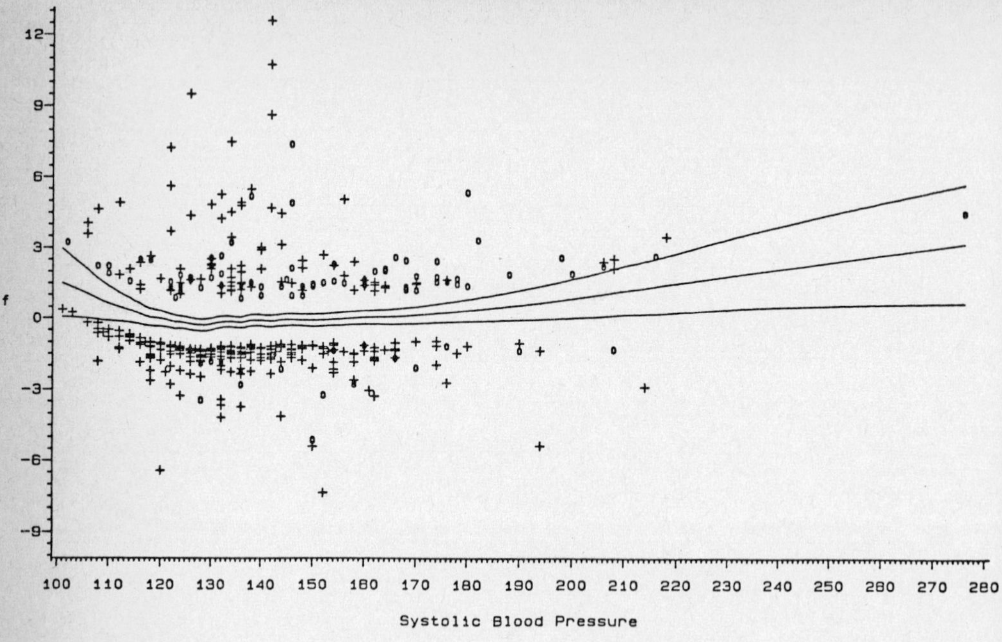

Figure 4.4b

Marginal fit for SBP conditioned on treatment

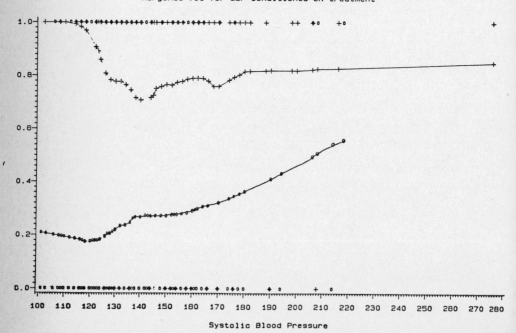

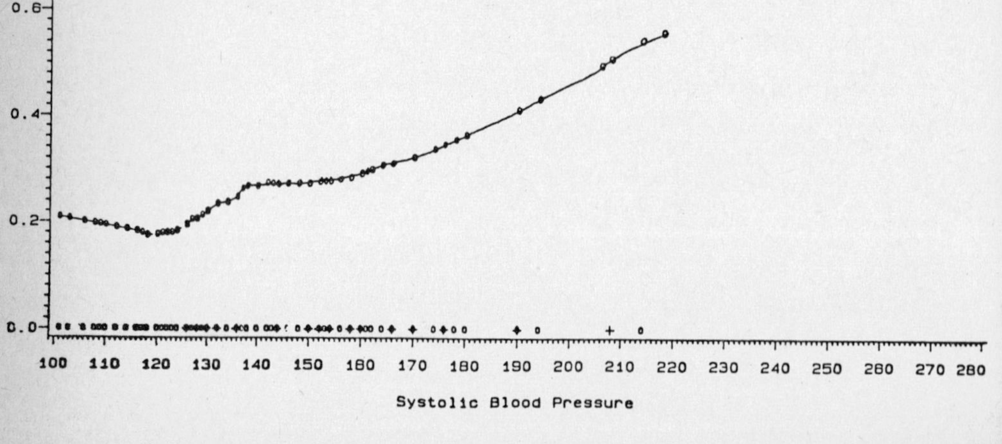

Figure 4.5a

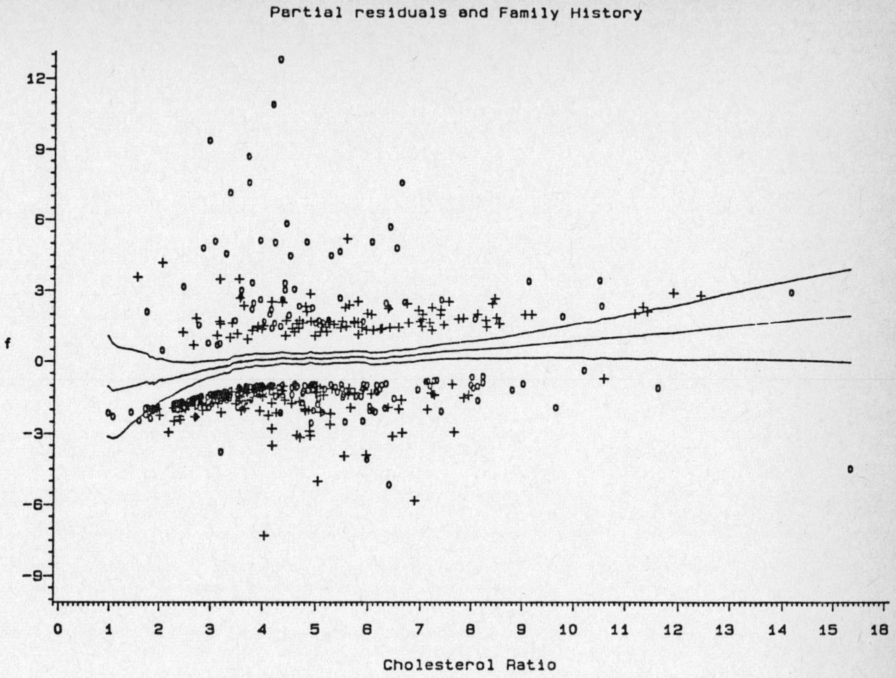

Figure 4.5b

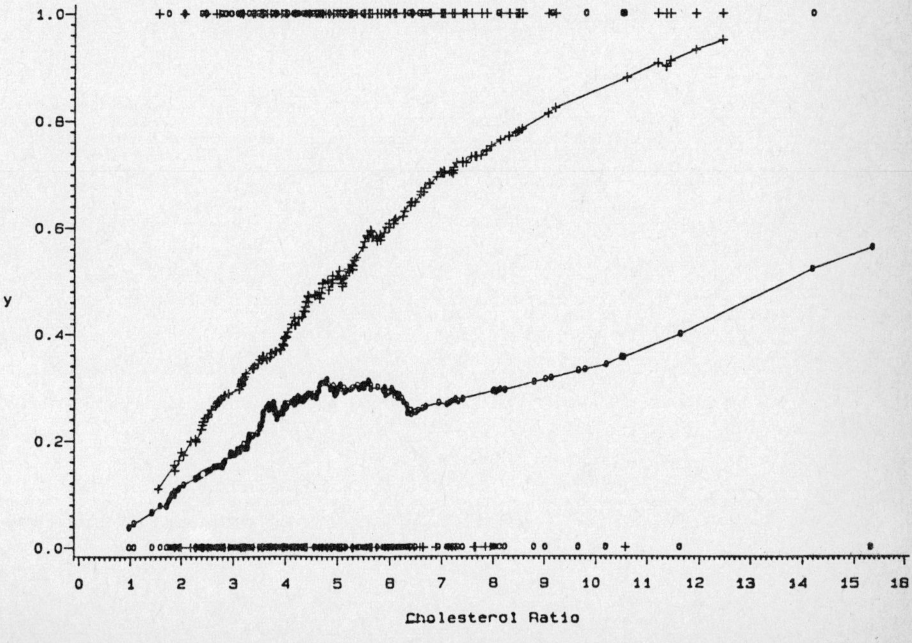

ON THE GUHA APPROACH TO MODEL SEARCH IN CONNECTION TO GENERALIZED LINEAR MODELS

Tomáš Havránek
Center of Biomathematics, Institute of Physiology,
Czechoslovak Academy of Sciences, Prague

Dan Pokorný
Department of Applied Mathematics, Psychiatric
Research Institute, Prague

SUMMARY

The GUHA (General Unary Hypotheses Automaton) method has been developed during last twenty years from a particular device for analysing simple categorial data to a rather general theoretical framework for construction of data analytic programs. The approach is primarily based on the logic of finite data structures (finite models in the usual logical sense) using deductive power of some calculi for a reduction of the search over a set of possible hypotheses (models in the statistical sense). Here, first, some background is discussed and illustrated by simple examples, then some suggestions for model search methods over generalized linear models are explained and, finally, statistical questions concerning the coherence of statistical decision rules and related problems are briefly mentioned.

Keywords: MODEL SEARCH; CONTINGENCY TABLES; LOG-LINEAR MODELS; FACTORIAL MODELS; GLIM; COHERENT STATISTICAL PROCEDURES; HYPOTHESIS GENERATION.

1. INTRODUCTION

First, let us mention that we shall use, where possible, the usual statistical terminology. Hence models are hypotheses in the statistical sense. Current literature on GUHA methods, e.g. Havránek and Hájek (1978a,b) use logical terminology and this could be sometimes a source of misunderstanding.

As McCullagh and Nelder (1983) point out, data often support with almost equal emphasis several possible models and it is important that the analyst accept this. Our experience is that in practice, wrongly, only one best model is searched for. On the other hand, the GUHA procedures were build to offer the analyst a set of all models (from a class of models in question) supported by data and ordered by some simplicity relation. Clearly such a search over a class of possible models must be in sense optimized to become computationally tractable.

We shall consider, under some frame assumptions, a set of statistical models. This set will be denoted RQ (relevant question). Two examples: (i) frame assumptions: $\underline{Y}=\underline{X}\beta+\underline{e}$, where $\underline{X}$ is a given matrix of reals (nXp), β is a vector of real parameters and $\underline{e}$ is a random vector having $N(\underline{0},\sigma^2 \underline{I})$ distribution. RQ = $\{(d_1,\ldots,d_p) \in \{0,1\}^p\}$ with interpretation: if $d_i = 0$ then $\beta_i = 0$. (ii) The independent multinomial sampling scheme with p $\{0,1\}$-valued variables. RQ = {all hierarchical log-linear models} (or RQ = {all hierarchical log-linear models containing some prescribed interactions}).

The task is now to find all models from RQ that are supported by a data set $\underline{M}$ (from a class $\mathcal{M}$ of all possible data sets under frame assumptions) and, in addition, to find in a sense the strongest or more interesting of them. Two things are necessary for completing such a task (defined by RQ and $\mathcal{M}$): First, to have a *syntactical description* of RQ, that is equivalent, in fact, to have an algorithm for generating all models in RQ. Second, to have a link between each $m \in$ RQ and each $\underline{M} \in \mathcal{M}$ giving a decision whether m is supported by $\underline{M}$ or not.

For a syntactical description think e.g. RQ to be a set of all graphical models in the multinomial case. For a link, think a goodness-of-fit test d on a given α-level, i.e., $d(m,\underline{M},\alpha) \in$ {reject, accept} = {r,a}. Clearly, here is an asymmetry in interpreting both decisions. Some other decision rules could be considered, e.g. a normalized deviance greater than a given limit.

2. ORDERINGS ON A SET OF MODELS

A set of models RQ is partially ordered by semantical embedding of models: $m_1 \models m_2$ (read: if m_1 is true, the m_2 is true). We speak now on *theoretical level*, hence "true" refers to probability structures, parametric spaces etc. (a logically clear semantics was developed in Hájek and Havránek, 1978). Sometimes it could be hard to decide whether $m_1 \models m_2$ or not. But in reasonable cases we have a (partial) ordering $\leq$ on RQ such that we are able to see *at first glance* whether $m_1 \leq m_2$ or not and, moreover, such that $m_1 \leq m_2$ implies $m_1 \models m_2$. Usually, $\leq$ is interpreted as a simplicity relation, i.e. if $m_1 \leq m_2$ and $m_1 \neq m_2$ we say that m_1 is simpler than m_2.

Examples: In the above linear case (for p=4), $m_1=(1,0,1,0) \leq m_2=(1,1,1,0)$. For hierarchical log-linear models in the generating class notation: m_1=(AB,CD) $\leq m_2$ = =(ABC,CD). In the GLIM notation, e.g. 1+X+Z $\leq$ 1+X+Y+Z, A+B+C+D+A.B+A.C $\leq$ A+B+C+D+A.B+A.C+B.C+A.B.C+C.D.

2.1 *The simplicity relation in model search inquiry*

First, we can use two principles:

(i) if $m_1 \leq m_2$ and m_1 is accepted, then m_2 is to be accepted, and

(ii) if $m_1 \leq m_2$ and m_2 is rejected, then m_1 is to be rejected.

In both cases, m_2 or m_1, respectively, is not necessarily evaluated at the given data set $\underline{M}$. Cf. discussions in Edwards and Havránek (1985 a,b)

Second, if $A(\underline{M}) = RQ$ is a set of models supported by $\underline{M}$ (*accepted* models), we can find *minimal* elements of $A(\underline{M})$ with respect to $\leq$. If MIN(A) is such a set, we see at first glance that each $m \in A$ is to be accepted, using the knowledge of MIN(A) only. The notion of MIN(A) corresponds, in fact, to the notion of a solution with respect to RQ and $\underline{M}$ used in Hájek and Havránek (1978a,b).

Two comments: (a) For many reasons, statisticians do not consider both rules (i) and (ii) to have the same validity. For example, Cox and Snell (1974) use (i) only, Havránek (1984) uses (ii) only and Edwards and Havránek (1985a,b) use both. (b) From the logical point of view, the relation $\leq$ corresponds to a deduction (a proof). If $m_1 \leq m_2$ then m_2 can be deduced from m_1 ($m_1 \vdash m_2$). Since we demand $\leq$ to be easily recognisable, we can view $\leq$ as a deduction rule, i.e. one-step deduction. As $\leq$ defined above, this deduction rule is *sound*, i.e. if $m_1 \leq m_2$, then $m_1 \models m_2$. We can ask, whether $\leq$ is complete, i.e. whether $m_1 \leq m_2$ (or $m_1 \vdash m_2$) is equivalent to $m_1 \models m_2$. Clearly, complete relation $\leq$ is more powerful than incomplete relation in constructing $A(\underline{M})$ and MIN(A). For discussing such questions, it is useful to view models as formulas of some logical calculus.

2.2 *Coherence*

Even if we use both rules (i) and (ii), we still need a decision rule d: $RQ \times \mathcal{M} \to \{a,r\}$. Gabriel (1969) points out that a reasonable d should be coherent, i.e. if $m_1 \leq m_2$ and $d(m_2,\underline{M})=r$ then it should be $d(m_1,\underline{M})=r$. Or equivalently, $m_1 \leq m_2$ and $d(m_1,\underline{M})=a$ should imply $d(m_2,\underline{M})=a$. Many of frequently used decision rules are only approximately coherent, i.e. the case $m_1 \leq m_2$, $d(m_1,\underline{M})=a$, $d(m_2,\underline{M})=r$ can occur, but we hope that with a small probability (see 5.1 here). In some cases, decision rules can be changed, with a good interpretation, to be coherent (see 4.2).

3. ORDERINGS ON DATA

In the classical GUHA procedures another approach is frequently used. Consider now A,B,C,... to be {0,1}-valued variables. Let $\underline{M}$ be a data set containing values of these variables (and perhaps of some other variables) on some set of objects M. An *elementary conjunction*, e.g. κ = A&∼B&E, defines a subset of data (in fact a substructure) on M_κ containing only objects for which the value of A is 1, the value of B is 0, the value of C is 1 (∼ is here a negation sign). We say that an elementary conjunction κ_1 is a subconjunction of κ_2 ($\kappa_1 \subseteq \kappa_2$) if all variables contained in κ_1 are in κ_2 with the same signs.

If now, for a given data set $\underline{M}$, κ_1 and κ_2 define the same subset of objects (κ_1 and κ_2 are equivalent on $\underline{M}$, $\kappa_1 \equiv \kappa_2$) then each κ, $\kappa_1 \subseteq \kappa \subseteq \kappa_2$, defines the same subset and a value of every statistic computed on $\underline{M}_{\kappa_1}$ for some other variables is the same for $\underline{M}_\kappa$. Everything what we wish to know about $\underline{M}_\kappa$ could be deduced from our knowledge of $\underline{M}_{\kappa_1}$. It is important to note that for a given κ_1 such a maximal κ_2 (containing some of A,B,C ...) and equivalent on $\underline{M}$ with κ_1 could be easily found (Hájek and Havránek, 1978, Theorem 6.3.14). What is remarkable that another relations could be defined between elementary conjunctions (and hence between data sets) enabling such a deduction but only for some particular classes of statistics (see Hájek and Havránek, 1978a,b).

3.1 *An example: deduction with no theoretical counterpart*

Consider now the following task (a particular version of the implication GUHA procedure). A,B,C,...,P are {0,1}-valued covariates, Y is a {0,1}-valued response variable. Let $P(Y=1/\kappa)$ be the probability of Y=1 under conditions specified by κ (presence or absence of some covariates - properties). Define RQ = $\{\kappa;\ P(Y=1/\kappa) \geq p_0$, κ is an elementary conjunction of some covariates$\}$, where p_0 is a given real constant. Whichever decision rule for accepting $P(Y-1/\kappa) \geq p_0$ is applied, we can use, in the search for $A(\underline{M})$, the following trick: if, for a κ_1,$d(\underline{M},\kappa_1)$=a (or $d(\underline{M},\kappa_1)$=r) then find the greatest κ_2, $\kappa_1 \subseteq \kappa_2$ and $\kappa_1 \equiv \kappa_2$ for $\underline{M}$, and deduce (without computation) that $d(\underline{M},\kappa)$=a (or $d(\underline{M},\kappa)$=r) for each κ, $\kappa_1 \subseteq \kappa \subseteq \kappa_2$. Clearly, as a result we shall print only κ_1, κ_2 and $d(\underline{M},\kappa_1)$=a (or =r).

This example is rather primitive, but it illustrates one substantial fact - there is a deduction on the data level that should be useful, but it has no reasonable counterpart in our theoretical (model) level.

3.2 *Pseudo-deduction*

Sometimes the situation can be a little more complicated. Consider again {0,1}-valued covariates A,B,...,P and two, say continuous, variables Y_1, Y_2. Let RQ = $\{\kappa;$ Y_1 and Y_2 are positively dependent under $\kappa\}$. Denote $n(\underline{M},\kappa)$ the number of objects in $\underline{M}$ for which κ is satisfied. We shall use a rank correlation coefficient. Let $R(Y_1,Y_2,\kappa)$ be its value on $\underline{M}_\kappa$. If now $R(Y_1,Y_2,\kappa_1) = 1$ then $R(Y_1,Y_2,\kappa_2) = 1$ for each κ_2, $\kappa_1 \subseteq \kappa_2$ (subject to $n(\underline{M},\kappa_2) \geq 2$). If now α is a given significance level, we can define

$A(\underline{M}) = \{\kappa;\ R(Y_1,Y_2,\kappa) \geq r(\alpha,\ n(\underline{M},\kappa))\}$, where $r(\alpha,n)$ is a critical level. Note that acceptance here means the rejection of the classical null hypothesis of independence.

If now we find that $R(Y_1,Y_2,\kappa_1) \geq r(\alpha,n(\underline{M},\kappa_1))$, it is worth to ask whether $R(Y_1,Y_2,\kappa_1)=1$. If yes, then we can deduce for each $\kappa_2, \kappa_1 \subseteq \kappa_2$ that $R(Y_1,Y_2,\kappa_2)$ is significant subject to $n(\underline{M},\kappa_2) \geq n_o(\alpha)$, where $n_o(\alpha)$ is the minimal number of cases for which R can be significant for the given α-level. The complication is here that we need to evaluate $n(\underline{M},\kappa)$.

But, in fact, we can use a pseudo-deduction: if $R(Y_1,Y_2,\kappa_1)=1$ then evaluation of $R(Y_1,Y_2,\kappa_2)$ for any κ_2, $\kappa_1 \subseteq \kappa_2$, is *uninteresting* and a computer procedure should skip all such κ_2 (note that elementary conjunctions have to be generated in some linear ordering, hence we jump over a segment in this ordering; see Havránek and Pokorný, 1978). Has this deduction a theoretical counterpart? Is it reasonable that if $R(Y_1,Y_2,\kappa_1)=1$ then for all κ_2, $\kappa_1 \subseteq \kappa_2$, Y_1 and Y_2 are positively dependent?

3.3 *More on simplicity relations*

Let us summarize that now a model search inquiry is given by a partially ordered class of possible models RQ and a class of possible data sets M. Up to now, we assumed that the relation is a semantically based simplicity relation ($m_1 \leq m_2$ implies $m_1 \models m_2$). An example of not semantically based simplicity relation (where soundness has no sense): In 3.1. the conjunctions can be ordered by their length (the number of variables contained).

A computer procedure for finding $A(\underline{M}) \subseteq RQ$, should generate members of RQ and evaluate $d(m,\underline{M})$ for them. Clearly some deduction or various tricks can be used to avoid computing $d(m,\underline{M})$ for each $m \in RQ$. Effectiveness of such tools depends on the way in which members of RQ are generated. There has to be some linear ordering $\underline{\leftarrow}$ defining such a sequence. There are various design choices which have to be made in constructing such a procedure:

(a) Should $\underline{\leftarrow}$ be given apriori or should it be defined successively in dependence on evaluation of some models?

(b) Should $\underline{\leftarrow}$ be consistent with $\leq$?

In classical GUHA procedures $\underline{\leftarrow}$ is an apriori given ordering and embeds $\leq$ which is generally not semantically based. Using a deduction rule on the data level, in particular runs of a procedure, some segments of $\underline{\leftarrow}$ are overjumped. In Havránek (1984), for graphical models, $\underline{\leftarrow}$ is apriori given and embeds semantically based simplicity relation, but in an opposite direction, i.e. if $m_1 \leq m_2$ then $m_2 \underline{\leftarrow} m_1$ (only the rule (ii) from 2.1 is used). In Edwards and Havránek (1985a,b) $\underline{\leftarrow}$ is not apriori given and is not generally consistent with $\leq$. But the case in which $\leq$ entails $\underline{\leftarrow}$ is covered.

We feel that it is necessary to discuss model search programs systematically from this point of view. On the other hand, it is not generally clear whether really

$\leq$ should entail $\underline{\leftarrow}$. It is worth to mention that here a contradiction with another point of view can occur, namely with complexity of procedures, measured e.g. by the number of evaluated decisions $d(m,\underline{M})$ to obtain a comprehensive description of $A(\underline{M})$.

4. GLIM APPLICATIONS

Using the ideas discussed above, some model search procedures over GLIM can be suggested. We are aware that model search by GLIM was discussed by many authors. Some systematic procedures were presented e.g. by Whittaker (1982) and Deischel (1983). We hope that stressing some facets of such procedures can lead to some new, perhaps useful, views.

4.1 *Factorial models*

Consider models in GLIM (see the section 3.4 of McCullagh and Nelder, 1983). Basically, we have variables (primitive terms) A,B,C... and connectives + and . . Formulas are defined as follows: (i) A,B,C... are formulas, (ii) if ϕ and ψ are formulas then $\phi + \psi$ is a formula, (iii) if ϕ and ψ are formulas, then $\phi . \psi$ is a formula.

There is a convention: + has a lower priority than . (i.e. $\phi \cdot \psi + \chi \equiv (\phi \cdot \psi)\chi$). We assume such a semantics that the following theorems hold:

$$\begin{aligned}
\phi \cdot \psi &\equiv \psi \cdot \phi \,, \\
\phi + \psi &\equiv \psi + \phi \,, \\
(\phi \cdot \psi) \cdot \chi &\equiv \phi \cdot (\psi \cdot \chi), \\
(\phi + \psi) + \chi &\equiv \phi + (\psi + \chi), \\
(\phi + \psi) \cdot \chi &\equiv \phi \cdot \chi + \psi \cdot \chi
\end{aligned} \tag{1}$$

We can define a normal form: $\phi_1 + \ldots + \phi_k$, where ϕ_i are distinct and each ϕ_i is of the form $\psi_{i_1} . \ldots . \psi_{i_l}$, where $\psi_{i_1},\ldots,\psi_{i_l}$ are distinct variables. Each formula ϕ has, due (1), an equivalent normal form NF(ϕ). Simplicity ordering can be defined by $\phi \leq \psi$ if NF(ϕ) $_=$ NF(ψ) as a subformula). Note that connectives * and / are definable: $\phi * \psi = \phi + \psi + \phi \cdot \psi$, $\phi / \psi = \phi + pt(\phi) \cdot \psi$. Formulas containing these connectives can be transformed to normal forms too. The ordering $\leq$ is sound. If we consider hierarchical models (see 4.3), then it is complete (but this answer is not so trivial how it looks at first glance).

4.2 *A decision rule - coherence*

Suppose now that in our model search inquiry factors A,B,...,P are given. Then we can consider a saturated model I, defining a matrix $\underline{X}_I$. The model I contains all factors and interactions upt to A. ... .P. Suppose, moreover, that $\underline{X}_I$ is an nXr matrix with n > r, i.e. our data contains n > r objects, and, in additions let $\underline{X}_I$ is of full rank. Under standard linear model assumptions with normal residuals, we can use a goodness of-fit statistic

$$F(\phi) = \frac{(\hat{\mu} - \hat{\mu}(\phi))'(\hat{\mu} - \hat{\mu}(\phi))/(r - r(\phi))}{(\underline{Y} - \hat{\mu})'(Y - \hat{\mu})/(n - r)} \qquad (2)$$

where $\hat{\mu}$ and $\hat{\mu}(\phi)$ are estimates under $\underline{X}_I$ and ϕ respectively. For a given significance level α, the model ϕ is rejected if $F(\phi) \geq f_{\alpha,\phi}$, where $f_{\alpha,\phi} = F_{1-\alpha}(r-r(\phi),n-r)$, i.e. the $1-\alpha$ quantile of the Fisher distribution with $r-r(\phi)$ and n-r degrees of freedom. But this rule is not coherent. Using the same rationale as in Aitkin (1974) and Edwards and Havránek (1985b), this decision rule can be changed to be coherent using $f'_{\alpha,\phi} = r(r-r(\phi))^{-1} F_{1-\alpha}(r,n-r)$. This procedure has good simultaneous inference properties. For each ϕ it is a conservative test procedure, since $f'_{\alpha,\phi} \geq f_{\alpha,\phi}$.

4.3 *Model search - lattice properties*

Consider hierarchical models: a model ϕ is hierarchical if for each subformula $\psi_1 \cdot \psi_2$ of NF(ϕ), $\psi_1 + \psi_2$ is a subformula of NF(ϕ). RQ is now the set of all hierarchical models (using factors A,B,...,P). This set is a distributive lattice with respect to $\leq$, with join + and meet defined by intersection of generating classes (of normal forms). The paralelism with hierarchical log-linear models is clear, hence the procedure suggested by Edwards and Havránek (1985a,b) could be applied. This procedure uses fully the deductive power of $\leq$.

Clearly, in experimental or clinical research the need for such a models search is rare; for small number of factors the situation could be solved by fitting all models and testing differences. On the other hand, in large observational studies or survey data, such simple procedures could be useful.

The situation is fully analogous, if there is a given nesting model ϕ containing some factor variables and we consider models of the form ϕ/ψ where ψ is a model formula containing some other variables. If ψ can be any hierarchical model containing some variables not in ϕ, then we obtain again a distributive lattice.

Serious complications occur, if $\underline{X}_I$ is not of the full rank or $n \leq r$. In such a situation it has no sense to consider some models and the decision rule remains unclear. Generally, in such a situation, a procedure respecting the ordering $\leq$ in generating models to be evaluated is preferable. But when here a model ϕ is accepted? Perhaps if: (i) it can be tested and (ii) $\underline{\beta}_\phi = 0$ is rejected and ((iii) it has no testable succesors in $\leq$ or (iv) no of its testable succesors give significantly better fit).

4.4 *Continuous covariates*

Consider now the case where no compound terms involving more than one continuous covariate are allowed in a model formula. For simplicity, consider no power member (i.e. no X.X or X.Y) and factors A,B,C and continuous covariates X and Y only. If we use all hierarchical models, we obtain again a lattice structure with joint irreducible elements A,B,C,X and Y and meet irreducible elements ϕ = (ABX,ACX,BCX,ABCY) and ψ = (ABCX,ABY,ACY,BCY) using the generating class notation, i.e. NF(ϕ) contains all subformulas of A.B.X, A.C.X, B.C.X and A.B.C.A as additive elements. Note that the largest element is (ABCX,ABCY). Again the procedure of Edwards and Havránek (1985a,b) can be used.

4.5 *GLIM as a software tool for mechanized model search*

During the theoretical development of procedures for mechanized model search we definitely needed a software tool for "simulated" procedure runs. GLIM offers here some possibilities: (i) simulated run of a "mechanized" model search procedure can be performed within just one GLIM session, (ii) it is possible to make steps between near models by adding and/or deleting effects and (iii) the operations same for all models (such as result displaying) can be defined as a macro. Hence a "mechanized" search can be done by a user that, in accordance with the given algorithm, controls the selection of models to be tested with the help of GLIM. This way is sometimes tractable (in lattice structures) but it can be recommended only to a procedure developer or a very enthusiastic user.

On the other hand, GLIM does not allow full implementation of search procedures. For this purposes some extensions of the GLIM syntax are needed: (i) *Data structures:* variables of the type model, both scalars and arrays, with appropriate operations like join, meet etc., and (ii) *Control mechanisms,* allowing loops like e.g. conditional and unconditional jumps to an arbitrary directive within the GLIM program.

5. TWO STATISTICAL QUESTIONS

5.1 *Approximate coherence*

As we said, some decision rules are approximately coherent. In fact we have only some experience with real data sets that supports this belief. Since we started our constructions with non-coherent procedures for hierarchical log-linear case, we shall consider now likelihood ratio tests for this case (under multinomial independent sampling scheme).

Consider now the situation of three distinct models $m_0 \leqq m_1 \leqq m_2$. Let χ_i^2 is is the statistic for testing m_i, df_i the corresponding degrees of freedom and $\chi^2_{1/2}$ the statistic for testing m_1 with respect to m_2. Let c_i be critical levels for a given α-level. Suppose now that m_o is the true model. The non coherence error is $\chi^2_2 \geq c_2$ and $\chi^2_1 < c_1$ which is equivalent to the event

$$c_2 \leq \chi^2_2 , \qquad \chi^2_2 + \chi^2_{1/2} < c_1 \tag{3}$$

This probability of this even can be expressed by

$$P(c_2 \leq \chi^2_2 < c_1)\, P(\chi^2_{1/2} < c_1 - c_2) \tag{4}$$

Note that under m_o both χ^2_2 and $\chi^2_{1/2}$ have the central chi-square distribution with df_2 and $df_2 - df_1$. To obtain an idea about the magnitude of this probability, let us suppose that $df_1 = 20$, $\alpha = 0.05$ and present Table 1.

TABLE 1.

df_2	$P(c_2 \leq \chi^2_2 < c_1)$	$P(\chi^2_{1/2} < c_1 - c_2)$	$P((3))$
21	0.0133	0.7385	0.0098
22	0.0234	0.7155	0.0167
23	0.0308	0.7117	0.0219
24	0.0363	0.7132	0.0259
25	0.0402	0.7166	0.0288
26	0.0431	0.7209	0.0311
27	0.0452	0.7253	0.0328
28	0.0466	0.7298	0.0340
29	0.0477	0.7342	0.0350
30	0.0485	0.7385	0.0357
∞	0.05	0.95	0.0475

This situation is from the interpretational point of view the worst one: due to the errorneous rejection of m_2, m_1 is rejected.

Let now $df_2 = 20$, $\alpha=0.05$ and $P(\chi^2_2 \geq c_2) = 0.50$. Then χ^2_2 is non central with $= 12.262$. Consider further $P(\chi^2_{1/2} < c_1 - c_2)$ with noncentrality λ_1 increasing by 1 with each degree of freedom. We obtain Table 2.

TABLE 2.

$df_2 - df_1$	λ_1	$c_1 - c_2$	$P(\chi^2_{1/2} < c_1 - c_2)$	$P(c_2 \leq \chi^2_2 < c_1)$	$P((3))$
1	1	1.261	0.520	0.052	0.027
2	2	2.514	0.381	0.102	0.039
3	3	3.762	0.295	0.150	0.044
4	4	5.005	0.239	0.195	0.047
5	5	6.242	0.196	0.235	0.046
6	6	7.475	0.161	0.273	0.045
7	7	8.703	0.134	0.305	0.041
8	8	9.927	0.113	0.329	0.037
9	9	11.147	0.094	0.360	0.034
10	10	12.363	0.079	0.383	0.030

Are these probabilities so small that we can speak about approximate coherence?

5.2. *Comparing of non nested models*

The set MIN(A) contains non nested models (not ordered by ≤). Sometimes a non trivial preference relation among them can be constructed. In the GUHA procedure COLLAPS (Pokorný and Havránek, 1978, Havránek and Pokorný 1978 and Pokorný 1982) for an RXC contingency table models are defined by collapsing some rows and columns to obtain 2X2 tables. The procedure looks for sources of dependence in the original RXC table. Intensity of the dependence in various 2X2 tables can be compared by a test derived using the δ-method (see Havránek, 1980; unfortunately there is omitted by a typographical error the factor 1/m in the formula (5.12) for covariance). Some simulations indicate that this asymptotical test has reasonable properties for finite samples. If now two 2X2 tables are significantly distinguished by this test (one of them is better), we can say that one of them (the corresponding model) is in the preference relation before the second one.

In the multinomial case, we can use for comparing two models a statistic $D(r) = \Sigma_i\, r_i \log(\hat{p}_i^{m_2}/\hat{p}_i^{m_1})$, where r_i is the relative frequency in the cell i and $\hat{p}_i^{m_j}$ are estimates under m_j. If now, in a multidimensional table m_1 and m_2 are decomposable, we can principially use again the δ-method since the estimators are closed expressions depending on relative marginal frequencies. For example, in three dimensional table 2X2X2, for models m_1 = (AB,C) and m_2 = (A,BC) i.e. $p_{ijk} = p_{ij.}p_{..k}$ and $p_{ijk} = p_{i..}p_{.jk}$, the statistic $\sqrt{n}\,(D(r) - D(p))$ has asymptotically normal distribution with zero means and variance

$v(p) = \Sigma_{ijk}\, p_{ijk} d_{ijk} - \Sigma_{ijk}\Sigma_{i'j'k'}\, p_{ijk} d_{ijk} p_{i'j'k'} d_{i'j'k'}$, where

$d_{ijk} = \log(p_{i..}p_{.jk}/p_{ij.}p_{..k}) + 2(1/p_{.jk} + 1/p_{ij.}) + 4(1/p_{i..} + 1/p_{..k})$. Since, if models are equivalent then D(p)=0, we can use a test statistic $D(r)(n/v(r))^{1/2}$. Clearly, this way is very complicated. The only possibility for a practical application is to write a program, computing symbolically v(p) to given pairs of decomposable models (computing and using partial derivatives of D(p)) and then using this variance in evaluating the test statistic.

REFERENCES

AITKIN, M.A. (1974). Simultaneous inference and the choice of variable subsets in multiple regression. Technometrics, 16, 221-227.

COX, D.R. and SNELL, E.J. (1974). The choice of variables in observational studies. Appl. Stat., 23, 51-59.

DEISCHEL, G. (1983). (Log)-linear models and lattices. GLIM Newsletter 6.

EDWARDS, D. and HAVRÁNEK, T. (1985a). A fast procedure for model search in multidimensional contingency tables. Biometrika, 72 (in print).

EDWARDS, D. and HAVRÁNEK, T. (1985b). A fast model selection procedure (submitted).

GABRIEL, K.R. (1969). Simultaneous test procedures - some theory of multiple comparisons. Ann. Math. Statist., 40, 224-250.

HÁJEK, P., HAVEL, I. and CHYTIL, M. (1966). The GUHA method of automatic hypotheses determination. Computing, 1, 293-250.

HÁJEK, P. and HAVRÁNEK, T. (1977). On generation of inductive hypotheses. Int. J. Man-Machine Studies, 9, 415-438.

HÁJEK, P. and HAVRÁNEK, T. (1978a). Mechanizing hypothesis generation - mathematical foundations of a general theory. Springer-Verlag.

HÁJEK, P. and HAVRÁNEK, T. (1978b). The GUHA method - its aims and techniques. Int. J. Man-Machine Studies, 10, 3-22.

HAVRÁNEK, T. (1980). Some comments on GUHA procedures. Explorative Datenanalyse, Springer-Verlag, 156-177.

HAVRÁNEK, T. and POKORNÝ, D. (1978). GUHA-style processing of mixed data. Int. J. Man-Machine Studies, 10, 47-58.

McCULLAGH, P. and NELDER, J.A. (1983). Generalized linear models. Chapman and Hall.

POKORNÝ, D. (1982). Procedures for optimal collapsing of two-way contingency tables. COMPSTAT 1982 , Physica-Verlag, 96-102.

POKORNÝ D. and HAVRÁNEK, T. (1980). On some procedure for identifying sources of dependence in contingency tables. COMPSTAT 1980, Physica-Verlag, 221-227.

WHITTAKER, J. (1982). GLIM syntax and simultaneous tests for graphical log-linear models. GLIM 82: Proceedings of the international conference on generalized linear models, Lecture Notes in Statistics, vol. 14, Springer-Verlag.

ESTIMATION OF INTEROBSERVER VARIATION FOR ORDINAL RATING SCALES.

By Bent Jørgensen
Odense University, Denmark.

Summary

The paper introduces a log-linear model for interobserver variation for the case where several observers have rated the same objects on an ordinal scale with r categories. The model, which is based on assignment of scores to the categories, allows a distinction between random and systematic differences between observers, and allows estimation of the proportion of cases in which there is complete agreement between observers. The model is illustrated using a set of data on radiologic diagnosing of arthritis.

Key words: Categorical data; Interobserver agreement; Kappa; Log-linear models; Ordinal data.

1. Introduction.

Consider an experiment in which n objects have been classified on an ordinal scale with categories 1,...,r by each of k observers. We may summarize the results of the experiment as an r^k contingency table

$$y = \{y_i : i = (i_1,...,i_k);\ i_j = 1,...,r;\ j = 1,...,k\} ,$$

where y_i denotes the number of objects classified by observer j as belonging to group i_j, j=1,...,k. We have mainly medical applications in mind, and in the following we refer to the objects as "patients". We assume that the categories have been assigned scores $x(1)<x(2)<\cdots<x(r)$, which reflect, in some sense, the severity of the disease being diagnosed. We consider the problem of estimating the interobserver variation of the classification procedure, based on data of the above form.

Cohen (1960, 1968) suggested a measure of interobserver variation called kappa, which in its "weighted" form may be used with ordered categories. The kappa statistic is widely used in medical applications; a review of the literature on kappa is given by Fleiss (1981, Chapter 13). It is not clear, however, exactly what the kappa statistic measures, and for example Jensen (1983) showed that the value of kappa depends not only on the variation between observers, but also on the distribution of the patients over the r categories, making kappa unsuitable as a general measure of interobserver variation. This is a serious objection to the kappa statistic, but it must also be stressed that the calculation of a single index of agreement is a very crude form of statistical analysis, in which important features of the data are easily overlooked.

One alternative possibility is to treat the score x as a continuous variable, and use for example a normal variance component model. However, in the present paper we consider a model which takes the discreteness of the rating scale into account, being based on a log-linear analysis of the contingency table **y**. Tanner and Young (1985) suggested a log-linear model for analysing interobserver variation for a rating scale with unordered categories. In our model, the ordering of the scale is taken into account via the score x. In an approximate sense, the model treats x as a discretization of an underlying normally distributed variable.

The model is introduced in Section 2. In Section 3 we consider the interpretation of the model, and we consider hypothesis testing within the model. In Section 4 we analyse a set of data on the radiologic evaluation of arthritis of the hip.

2. The model.

Consider the contingency table **y** introduced in Section 1, and let $\mu_i = E(y_i)$ denote the expected number of observations in cell **i** = $(i_1,\ldots,i_k)$. By **x** = **x**(**i**) = $(x(i_1),\ldots,x(i_k)) \in \mathbb{R}^k$ we denote the vector of scores in cell **i**. Let

$$\bar{x}(\mathbf{i}) = \bar{x} = \frac{1}{k} \sum_{j=1}^{k} x(i_j)$$

denote the average score for cell **i**. If the scores $x(1),\ldots,x(r)$ are more or less regularly spaced, there is only a restricted number of distinct possible values for the average score $\bar{x}$. We let $g(\mathbf{i}) = g \in \{1,\ldots,\ell\}$ denote an index that labels the groups obtained by partitioning **y** according to the value of $\bar{x}$. Since the average score in a given group is constant, we may think of the patients in the group as having approximately the same true score.

For a given vector μ in $\mathbb{R}^k$, let $A(\mu)$ be the affine space defined by

$$A(\mu) = \mu + \{(t,\ldots,t)^T: t \in \mathbb{R}\} ,$$

and let p^{μ} denote the orthogonal projection onto $A(\mu)$, given by

$$p^{\mu}(\mathbf{x})_h = \mu_h + \frac{1}{k} \sum_{j=1}^{k} (x_j - \mu_j)$$

$$= \mu_h + \bar{x} - \bar{\mu} \qquad (h=1,\ldots,k)$$

where $\bar{\mu} = (\Sigma \mu_j)/k$. Our basic model for interobserver variation is defined by

$$\log \mu_{\mathbf{i}} = \alpha_g + \delta d_{\mathbf{i}} - \beta \frac{1}{2} \|\mathbf{x} - p^{\mu^{(g)}}(\mathbf{x})\|^2$$

$$= \alpha_g + \delta d_{\mathbf{i}} - \beta \frac{1}{2} \|\mathbf{x} - \bar{x}\mathbf{e} - (\mu^{(g)} - \bar{\mu}^{(g)}\mathbf{e})\|^2 \qquad (2.1)$$

where

$$\mathbf{d}_i = \begin{cases} 1 & \text{for } i_1 = \cdots = i_k \\ 0 & \text{otherwise} \end{cases}$$

and $\mathbf{e} = (1,\ldots,1)^T$. The parameters of the model are $\delta \in \mathbb{R}$, $\beta \in \mathbb{R}_+$, $\mu^{(g)} \in \mathbb{R}^k$ and $\alpha_g \in \mathbb{R}$ for $g=1,\ldots,\ell$.

Using the expression

$$\frac{1}{2}\|\mathbf{x} - p^{\mu^{(g)}}(\mathbf{x})\|^2 = \frac{1}{2}\sum_{j=1}^{k}(x_j - \bar{x} - \mu_j^{(g)} + \bar{\mu}^{(g)})^2$$

$$= \frac{1}{2}\sum_{j=1}^{k}(x_j - \bar{x})^2 + \frac{1}{2}\sum_{j=1}^{k}(\mu_j^{(g)} - \bar{\mu}^{(g)})^2 - \sum_{j=1}^{k}\mu_j^{(g)}(x_j - \bar{x}),$$

where we have used $\Sigma(x_j - \bar{x}) = 0$, we may write (2.1) in the form

$$\log \mu_i = \tilde{\alpha}_g + \delta d_i - \beta\frac{1}{2}\sum_{j=1}^{k}(x_j - \bar{x})^2 + \sum_{j=1}^{k}\tilde{\mu}_j^{(g)}(x_j - \bar{x}), \qquad (2.2)$$

where $\tilde{\alpha}_g = \alpha_g - \frac{1}{2}\beta\Sigma(\mu_j^{(g)} - \bar{\mu}^{(g)})^2$ and $\tilde{\mu}_j^{(g)} = \beta\,\mu_j^{(g)}$. The model (2.1) is thus a log-linear model, which may be fitted using GLIM.

3. Interpretation of the model.

3.1 Two observers.

In the case of two observers ($k=2$), we define $z = z(i) = (x_1 - x_2)/\sqrt{2}$ and $\xi_g = (\mu_1^{(g)} - \mu_2^{(g)})/\sqrt{2}$, where z represents the discrepancy between the classifications by the two observers. We may, in this case, write (2.1) in the form

$$\mu_i = a_g \exp\{\delta d_i - \frac{1}{2}\beta(z - \xi_g)^2\}, \qquad (3.1)$$

where $a_g = \exp(\alpha_g)$. For a given group g, (3.1) is a normal probability density function with expected value ξ_g and variance $1/\beta$, with the norming constant $(\beta/2\pi)^{\frac{1}{2}}$ being absorbed into a_g. The factor a_g represents the number of patients in group g. The factor $\exp(\delta d_i)$, which is 1 for $i_1 \neq i_2$, represents for $\delta>0$ ($\delta<0$) a surplus (deficiency) of cases in which the two observers agree, relative to what is expected from the (approximately) normal distribution of z. This factor thus represents the cases in which there is no doubt about the classification. In the remaining cases, where there is doubt, the discrepancy z between the two observers is approximately normally distributed with expected value ξ_g. If ξ_g depends on g, there is thus complete <u>disagreement between observers</u> concerning the doubtful cases. If ξ_g does not depend on g, there is a constant systematic difference between the observers, and in this case we speak of <u>consistency between observers</u>. Finally, if the common value of ξ_g is zero, we speak of <u>agreement between observers</u>, meaning absence of any systematic differences between the observers.

In principle, β may be negative in (3.1) or (2.1), but if this happens, the distribution of z becomes bimodal with modes at the extremes of the range of z. This possibility thus represents an extreme degree of disagreement between the observers, and is unlikely in practice, at least for well-behaved classification procedures.

3.2 The general case.

In the general case, the discrepancy z is replaced by the vector of discrepancies $\mathbf{x}-\bar{x}\mathbf{e}$, which by (2.1) has, for group g, approximately a (k-1)-dimensional spherical normal distribution with expected value $\mu^{(g)}-\bar{\mu}^{(g)}\mathbf{e}$. Hence we take $\tilde{\mu}_1^{(g)}=0$ in (2.2), and the estimable quantities in (2.2) are $\tilde{\mu}_2^{(g)},\ldots,\tilde{\mu}_k^{(g)}$, which via

$$\mu_j^{(g)} = \beta^{-1}\tilde{\mu}_j^{(g)} \quad , \quad j=2,\ldots,k$$

represent the levels of observer 2,...,k, relative to observer 1, for group g. The parameters α_g and δ have the same interpretation as in the case k=2.

Just as in the case of two observers, the model (2.1) represents the possibility of disagreement between observers. The hypothesis of consistency between observers generalizes to

$$H_1: \mu^{(1)} = \cdots = \mu^{(\ell)} = \mu ,$$

which corresponds to $\tilde{\mu}^{(1)} = \cdots = \tilde{\mu}^{(\ell)}$ in (2.2). Hence this hypothesis is log-linear. The hypothesis of agreement between observers generalizes to

$$H_2: \mu = 0 ,$$

which is also a log-linear submodel of (2.2). The model (2.1) is denoted H_o.

The degrees of freedom for the hypotheses H_o, H_1 and H_2 depend on the way the scores x(1),...,x(r) are defined. In the simplest case of equally spaced scores, the degrees of freedom for the three hypotheses are

$$H_o: r^k - k\{k(r-1)-k\} - 4$$

$$H_1: r^k - kr - 2$$

$$H_2: r^k - k(r-1) - 3.$$

For r=2, the minimum number of observers required to obtain positive degrees of freedom for H_o,H_1 and H_2 are respectively k=5, 4 and 3. For r=3, H_o requires $k \geq 3$. In all other cases $k \geq 2$ is necessary to obtain positive degrees of freedom. It is possible to increase the degrees of freedom by letting the α_g's follow a suitable function depending on a few unknown parameters. Depending on the available information concerning the distribution of the patients over the categories, one can let the α_g's follow a smooth function, such as a normal probability density function, or one can use a piecewise constant function.

Under H_2, the estimates $\hat{\delta}$ and $\hat{\beta}$ represent the overall reliability of the classification procedure, according to the interpretation of δ and β given above. Under H_1, we need in addition to $\hat{\delta}$ and $\hat{\beta}$ the estimates $\hat{\mu}_1, \hat{\mu}_2, \ldots, \hat{\mu}_k$, where $\hat{\mu}_1=0$. If we may think of

$\mu_1,\ldots,\mu_k$ as having been drawn from a normal population, a suitable summary of the distribution of the μ_j's is

$$s^2 = \frac{1}{k-1} \sum_{j=1}^{k} (\hat{\mu}_j - \hat{\bar{\mu}})^2 \qquad (3.2)$$

which estimates the dispersion of the systematic differences between the observers.

4. A data example.

4.1 The data.

The data in the Appendix are the result of an experiment conducted in order to compare four different methods for assessing the severity of arthritis of the hip, based on X-ray pictures. Three of the methods (Wroblewski and Charnley, Hermodsson, and Heripret) involve scales with four categories, whereas the Keelgren & Lawrence scale has five categories. The Heripret scale is a modification of the original 0-10 scale, see Kjærsgaard-Andersen *et al.* (1985) for details. For the scales with four categories we use the scores x=0,1,2,3, whereas in the Keelgren and Lawrence scale with categories 0,1,2,3,4 we use x=0,3/4,6/4,9/4,3. In this way the scores have the same range for all four methods, making a direct comparison of the parameter estimates possible.

In the experiment, X-ray pictures from n=100 patients were evaluated by each of k=4 observers for each method, and the same 100 X-ray pictures were used for all four methods. The tables in the Appendix show the number of patients for each possible combination of the four observers' classifications.

4.2 Analysis of the data.

Estimates of the parameters in the model (2.2) were obtained using GLIM, and standard errors of derived parameters were obtained from the GLIM output using the δ-method. The data are extremely sparse outside the main diagonal of the table, and the χ^2 approximation to the distribution of the deviance is not applicable here. However, since the total number of observations (n=100) is reasonably large, the distribution of differences between deviances for nested models is approximately χ^2, and this may be used to test the hypotheses H_1 and H_2. The goodness of fit of the model was checked by inspection of the residuals from the fit, and no serious departures from the model were found.

Table 1

*Analysis of deviance for the four data sets. NS = not significant at 5% level, * = significant at 5% level, *** = significant at 0.1% level.*

		Deviance	DF	Δ Deviance	Δ DF
Wroblewski	H_o	45.74	208		
and	H_1	74.17	238	28.43^{NS}	30
Charnley	H_2	96.33	241	22.16^{***}	3
	H_o	56.19	208		
Hermodsson	H_1	91.94	238	35.75^{NS}	30
	H_2	100.4	241	8.46^{*}	3
	H_o	13.31	208		
Heripret	H_1	52.83	238	39.52^{NS}	30
	H_2	95.71	241	42.88^{***}	3
Keelgren	H_o	154.0	561		
and	H_1	198.3	603	44.3^{NS}	42
Lawrence	H_2	220.3	606	22.0^{***}	3

Table 1 shows the analysis of deviance for each of the four methods with tests of the hypotheses H_1 and H_2. For all four methods, H_1 was accepted and H_2 was rejected at the 5% level. The tables in the Appendix show the fitted values under H_1, and Table 2 shows estimates of the parameters of H_1.

Table 2

Parameter estimates under H_1 with standard errors shown in brackets.

	Wroblewski and Charnley	*Hermodsson*	*Heripret*	*Keelgren and Lawrence*
δ	2.62 (.74)	2.47 (.85)	7.49 (7.56)	.77 (.71)
$\beta^{-\frac{1}{2}}$	.61 (.05)	.62 (.05)	.46 (.04)	.70 (.05)
μ_B	.32 (.13)	.17 (.13)	.04 (.09)	-.53 (.16)
μ_C	.56 (.14)	.19 (.13)	-.49 (.11)	.14 (.16)
μ_D	.07 (.13)	.37 (.13)	-.08 (.09)	.02 (.15)
s	.26	.15	.24	.30

Taking the standard errors of the estimates as known and equal to their estimated values, we may test equality of the δ's, respectively β's, for the four methods, using the inverse variances as weights. The test statistic for equality of the δ's is SSD=4.48, which is not significant compared with a $\chi^2(3)$ distribution. The test statistic for equality of the β's is SSD=10.77, which is significant at the 2.5% level compared with a $\chi^3(3)$ distribution.

The table also shows the values of $\hat{\mu}$ and s, where s^2 is defined in (3.2). The four values of s^2 may be compared using Bartlett's test for homogeneity of variances. This yields the result B=1.11, which is not significant in a $\chi^2(3)$ distribution.

Based on the significant difference between the β's, the method with the largest value of β (smallest value of $1/\sqrt{\beta}$), namely Heripret, is formally the best method. The Heripret method also has the largest estimate of δ, indicating a high proportion of agreement between observers, but the estimate of δ is not significantly different from zero. Both Wroblewski and Charnley, and Hermodsson have $\hat{\delta}$ significantly different from zero, and the latter two methods have approximately the same value of $1/\sqrt{\beta}$. Based on the value of s, Hermodsson's scale is the better of the two. Keelgren and Lawrence seems to be the least reliable method in all three aspects, although no firm conclusions about δ and s are possible from the present data.

5. Discussion.

For data such as the above, with systematic differences between the observers, reliable estimates of the variation between the systematic levels of the observers require many observers in the study. This is a general feature of any statistical model for interobserver variation, a point which seems hardly to have been discussed in connection with the kappa statistic, where two observers is the rule.

If more observers are included in order to improve the estimate of the variation between the systematic levels, the table **y** becomes more sparse, so that more patients are required. More patients are also required to improve the precision of $\hat{\delta}$, perhaps in conjunction with a smooth model for the variation of the α_g's. The parameter β, however, is already quite precisely estimated from the present data.

For scales with many categories, such that the score x is approximately a continuous variable and a normal variance component model is appropriate, the analysis based on such a model is presumably not qualitatively different from our analysis, and moreover the sparseness of y is of no concern in such an analysis. However, a unique feature of the present model is the presence of the term $\exp(\delta d_i)$ which models the proportion of cases in which there is no doubt about the classification, a feature which is manifest in the data analysed above.

Acknowledgements.

I am grateful to Anders Mørup Jensen for helpful discussions, and to Per Kjærsgaard-Andersen, Finn Christensen, Niels Wisbech Pedersen and Steen Asmus Schmidt for presenting me with the problem and the data.

References.

Cohen, J. (1960). A coefficient of agreement for nominal scales. *Educ. Psychol. Meas.*, **20**, 37-46.

Cohen, J. (1968). Weighted kappa: Nominal scale agreement with provision for scaled disagreement or partial credit. *Psychol. Bull.* **70**, 213-220.

Fleiss, J. (1981). *Statistical Methods for Rates and Proportions*, 2nd ed. New York: Wiley.

Jensen, A.M. (1983). The interpretation of Cohen's kappa. Paper presented at the 4th International Meeting on Clinical Biostatistics, Paris, September 1983.

Kjærsgaard-Andersen, P., Christensen, F., Pedersen, N.W. and Schmidt, S.A. (1985). Interobservatør samt interskalær variation ved radiologisk graduering af hofteledsartrose. Unpublished manuscript.

Tanner, M.A. and Young, M.A. (1985). Modeling agreement among raters. *J. Amer. Statist. Assoc.*, **80**, 175-180.

Appendix.

Data from an experiment with four observers A,B,C and D. The table shows the number of patients for each possible combination of classifications by the four observers. The fitted values under H_1 are shown in italics. Only data from one method (Wroblewski and Charnley) are presented here.

Wroblewski and Charnley

		B = 0				B = 1				B = 2				B = 3			
		C															
A	D	0	1	2	3	0	1	2	3	0	1	2	3	0	1	2	3
0	0	12	4	0	1	3	1	0	0	0	1	0	0	0	1	0	0
		12	*4*	*.1*	*0*	*2*	*1*	*.9*	*0*	*0*	*.5*	*0*	*0*	*0*	*0*	*0*	*0*
	1	1	0	0	0	1	4	0	0	0	0	0	0	0	0	0	0
		1	*.6*	*.5*	*0*	*.3*	*3.7*	*.5*	*0*	*.1*	*.3*	*.2*	*0*	*0*	*0*	*0*	*0*
	2	0	0	0	0	0	0	0	0	0	0	0	0	0	0	0	0
		0	*.1*	*0*	*0*	*0*	*.1*	*.1*	*0*	*0*	*0*	*.1*	*0*	*0*	*0*	*0*	*0*
	3	0	0	0	0	0	0	0	0	0	0	0	0	0	0	0	0
		0	*0*	*0*	*0*	*0*	*0*	*0*	*0*	*0*	*0*	*0*	*0*	*0*	*0*	*0*	*0*
1	0	0	1	0	0	0	2	0	0	0	0	1	0	0	0	0	0
		.9	*.5*	*.4*	*0*	*.2*	*3*	*.4*	*0*	*.1*	*.2*	*.2*	*0*	*0*	*0*	*0*	*0*
	1	0	4	1	0	0	24	2	0	0	1	4	0	0	0	0	0
		.1	*1.5*	*.2*	*0*	*.8*	*23.7*	*1.5*	*.2*	*0*	*.8*	*1.7*	*0*	*0*	*0*	*0*	*0*
	2	0	0	0	0	0	0	0	0	0	0	1	0	0	0	0	0
		0	*0*	*0*	*0*	*0*	*.4*	*.9*	*0*	*0*	*.4*	*.3*	*.2*	*0*	*0*	*0*	*.3*
	3	0	0	0	0	0	0	0	0	0	0	0	0	0	0	0	0
		0	*0*	*0*	*0*	*0*	*0*	*0*	*0*	*0*	*0*	*0*	*.2*	*0*	*0*	*0*	*.1*
2	0	0	0	0	0	0	0	0	0	0	0	0	0	0	0	0	0
		0	*0*	*0*	*0*	*0*	*0*	*0*	*0*	*0*	*0*	*0*	*0*	*0*	*0*	*0*	*0*
	1	0	0	0	0	0	0	1	0	0	0	0	0	0	0	0	0
		0	*0*	*0*	*0*	*0*	*.3*	*.7*	*0*	*0*	*.4*	*.2*	*.1*	*0*	*0*	*0*	*.3*
	2	0	0	0	0	0	0	0	0	0	0	7	1	0	0	2	0
		0	*0*	*0*	*0*	*0*	*.2*	*.1*	*0*	*0*	*0*	*7.3*	*1.9*	*0*	*0*	*1*	*1.5*
	3	0	0	0	0	0	0	0	0	0	0	1	2	0	0	1	0
		0	*0*	*0*	*0*	*0*	*0*	*0*	*0*	*0*	*0*	*.5*	*.8*	*0*	*0*	*.4*	*.8*
3	0	0	0	0	0	0	0	0	0	0	0	0	0	0	0	0	0
		0	*0*	*0*	*0*	*0*	*0*	*0*	*0*	*0*	*0*	*0*	*0*	*0*	*0*	*0*	*0*
	1	0	0	0	0	0	0	0	0	0	0	0	0	0	0	0	0
		0	*0*	*0*	*0*	*0*	*0*	*0*	*0*	*0*	*0*	*0*	*.1*	*0*	*0*	*0*	*0*
	2	0	0	0	0	0	0	0	0	0	0	1	0	0	0	0	1
		0	*0*	*0*	*0*	*0*	*0*	*0*	*0*	*0*	*0*	*.4*	*.6*	*0*	*0*	*.3*	*.7*
	3	0	0	0	0	0	0	0	0	0	0	1	1	0	0	0	8
		0	*0*	*0*	*0*	*0*	*0*	*0*	*0*	*0*	*0*	*.2*	*.3*	*0*	*0*	*.2*	*8*

Genstat 5: a general-purpose interactive statistical package, with facilities for generalized linear models.

P.W. Lane and R.W. Payne
Statistics Department
Rothamsted Experimental Station
Harpenden, Herts. AL5 2JQ

SUMMARY

In version 5, Genstat has been redesigned to provide a simpler syntax, better programming facilities and convenient interactive use. Facilities for generalized linear models are an integral part of Genstat 5 and include, in particular, the provision of summaries of analysis, such as predictions formed from an individual model or analysis of deviance derived from a series of models. Other statistical facilities include analysis of variance, cluster and multivariate analysis, time series and non-linear regression models.

Keywords: generalized linear models; interactive computing; prediction; statistical languages.

1. Introduction

The potential usefulness of a general-purpose interactive statistical package, incorporating generalized linear models together with other statistical procedures, has long been recognized. Genstat provides such a range of facilities but the basic design of the versions, before Genstat 5, was done before interactive working became widely available. Thus, although releases 4.01 (1977) to 4.04 (1982) can be run interactively, they are not as convenient to use as packages intended specifically for interactive use. Much of the output is designed primarily for line printers - it can be printed on (narrower) terminals but may not look as attractive. Likewise, the default output from many directives overflows the screen of a video terminal. One aspect of the Genstat 4 command language is the distinction between the compilation and execution of blocks of commands. Although this adds to the efficiency of a batch program, it is inappropriate for interactive use, when commands need to be executed immediately after they are typed.

Thus a much revised version, Genstat 5, has been designed both to exploit changes in the computing environment since the last major revision (in 1973) and to take advantage of the experience gained during that period.

2. The design of Genstat 5

2.1 *Control structures and programming facilities*

The command language of Genstat (like that of GLIM) is more than just a means of invoking statistical procedures: it can be used as a programming language in its own right, enabling new forms of analysis to be developed, output to be customized, etc. Virtually any result that can be printed in an analysis can also be stored within Genstat, in a suitable data structure, for input to other analyses.

The macro structure, which fulfilled the role of subroutine in Genstat 4, is replaced in Genstat 5 by the *procedure*. Invocation of a procedure looks identical to the standard Genstat 5 directives, with its arguments being transferred via the option and parameter lists only. Procedures are accessed automatically as required: when Genstat 5 receives a command it first checks whether it is an instance of one of the standard directives, then whether it has the name of a procedure already in store and, failing that, it searches the various procedure libraries attached to the job. This will allow Genstat 5 to be "customized" for particular applications, and a library of useful procedures will be supplied with the program.

The implementation of labels and jumps in a command language requires whole blocks of commands to be compiled before execution (as in Genstat 4). In Genstat 5, jumps are replaced by IF-THEN-ELSE, CASE and EXIT, in accordance with current beliefs about good programming practice. Commands are usually compiled and executed one at a time. With loops, on the first pass, the commands are compiled, executed and remembered, one at a time; on subsequent passes, the stored form is executed. Further directives can request Genstat to pause, allowing the user to read and digest the current screenful of output.

High-quality graphics for plotters or graphics monitors, in Genstat 5, uses the NAG Graphics Supplement. This will make them easier to implement at particular sites.

2.2 *Data structures*

Scalars, variates, factors, matrices (rectangular, symmetric and diagonal) and tables (with and without margins) are available with only minor changes from Genstat 4. Similarly, structures to store individual identifiers, arithmetic expressions and model formulae are again provided.

The text vector no longer has any restriction on the number of characters in each unit (or line), and there are general facilities for text handling, such as concatenation and editing. Likewise, facilities for pointer vectors are much extended to allow their elements to be conveniently referred to, either by unit numbers or user-defined labels, or to allow the entire set of structures to which they point to be substituted into a list of structures. Pointers can have other pointers as their elements, and so the complete range of hierarchical data structures described by Lamacraft and Payne (1980) can be represented.

There are several compound data structures defined as standard, which have the same mode as the pointer, enabling their individual elements to be accessed. For example the components of the SSPM structure, involved in regression and multivariate analyses, are a symmetric matrix of sums of squares and products, a variate of means and a scalar storing the sum of the weights.

2.3 *Syntax*

During the last 12 years, new facilities have been added to Genstat mainly in an evolutionary manner. This policy has been convenient for existing users, whose programs have continued to work with new releases, but it eventually led to inconsistencies in syntax, directives with overlapping facilities and a general need for rationalization, which caused the language to become more difficult to learn and remember.

In Genstat 5, the syntax has been completely revised. The Genstat 4 syntax has a special character (single quote) at the start and end of directive names. Only one of these is necessary to distinguish directive names from identifiers of structures etc. In Genstat 5, we have chosen to have a symbol to mark the end of each command, as in MLP (Ross 1980), rather than at the start of a directive name, as in GLIM. For interactive work, it is more important to know that the end of a command has been reached (so that it can be executed), rather than to know that a new one is about to begin, and the GLIM3 requirement to type e.g. $FIT X $ is unnatural. The end-of-directive symbol is colon (:). By default, newline will also signify "end of directive", with back-slash (\) indicating continuation, although a command can be given to specify that newline be regarded as a space (as in Genstat 4). Ampersand (&) replaces the use made of colon, in Genstat 4, to mean "repeat last directive name and options".

All Genstat 5 directives have identical rules of syntax. The information required by a directive is specified by parameters and options. Parameters specify lists of arguments or settings which are to be dealt with in *parallel*, while options specify arguments or settings that are *global*, i.e. that operate for all the items in the lists of parameters. For example

```
ANOVA [PRINT=aov,means; WEIGHT=W] Y=Yield,Profit; FITTED=Fy,Fp
```

requests two analyses, one of the variate Yield, saving the fitted values in variate Fy, and the other of the variate Profit, saving the fitted values in variate Fp; for both these analyses, the analysis-of-variance table and tables of means are printed and the analysis is weighted as specified by the variate W. The option names "PRINT" and "WEIGHT" can be abbreviated to the minimum number of letters required to recognize them, taking the options in the order defined for the directive; alternatively, the option names with the succeeding equals sign may be omitted altogether if the options are specified in the defined order. Likewise parameter names, such as "Y" or "FITTED", can be abbreviated or omitted. The settings of options or parameters may be identifiers, textual strings, expressions, model terms or formulae, as appropriate for the directive concerned.

3. Facilities for Generalized Linear Models

Genstat has provided facilities for fitting generalized linear models since Release 3.09 in 1977. Standard models are available with Normal, Poisson, binomial, gamma or inverse Normal distributions, and with identity, log, logit, reciprocal, power, square-root, probit or complementary-log-log link functions. Other distributions and link functions can be dealt with by writing short procedures to carry out the necessary calculations to form the deviance and working variates, and invoking a procedure in the standard library to carry out the iterative fitting.

As in GLIM, generalized linear models are included as an extension of linear regression, so that all the facilities provided for classical regression models are also available for GLMs. In Genstat, these include directives for:

(1) fitting a specified model;
(2) modifying a model by adding, dropping or switching terms;
(3) displaying the effects of alternative single-term changes to a model;
(4) choosing the best change to a model, based on reduction in the mean deviance.

A TERMS directive is provided to list in advance all the terms to be considered in a sequence of models and identify the common set of units to be used - excluding missing values and taking account of restrictions in force. This directive is now optional before the FIT directive (as is the equivalent command in GLIM). The default output includes the current residual deviance and degrees of freedom; in Genstat 5, there is also a report on possible outliers and influential points. Other available output includes fully annotated tables of parameter estimates, with their standard errors, and fitted values, leverages and residuals. The residuals are deviance residuals standardized by their estimated variance. All results of the fitting process, such as the linear predictor and the variance-covariance matrix of the parameter estimates, can be extracted and stored in standard Genstat data structures.

A particularly useful feature of Genstat regression, linear or generalized linear, is the ability at any stage to display an accumulated summary of changes made to the model by successive commands. This contains the deviance and degrees of freedom associated with each term, in the order fitted, and the corresponding values for the residual at each stage.

Qualitative effects can be included in models in Genstat using model formulae, as developed for the ANOVA directive of Genstat and as used in GLIM. By default, conventions for parameterization of effects, are the same as in GLIM, but the full set of parameters for each term can be requested if required. A recently added feature in Genstat is the ability to produce tables which summarize effects in a fitted model. These are called *predictions* (Lane and Nelder 1982); they predict what the fitted values would have been for combinations of values of some of the variables, if the number of observations in the data had been balanced in a specified way with respect to the other variables in the model. Thus, a table can show the effect of one or two particular variables, adjusting for the effects of all the other variables - qualitative or quantitative. The method of adjustment uses sample proportions by default; alternatives are to enforce equal weighting of each level of a factor, or to apply some other predefined pattern of weights.

4. Other Statistical Facilities

4.1 *Analysis of variance*

Genstat contains a very general algorithm for the analysis of variance of designed experiments (Wilkinson 1970; Payne and Wilkinson 1977). All the commonly-occurring experimental designs can be analysed, including all completely randomized orthogonal designs, randomized block designs, Latin and Graeco-Latin squares, split plots, repeated-measures designs and lattices. Output includes the analysis-of-variance table, means with standard errors, and estimates for polynomial, and other, contrasts. Analysis of covariance is available with all these designs. Missing values are also catered for.

4.2 *Cluster Analysis*

Similarity matrices can be formed from either quantitative or qualitative variables and used for a range of hierarchical cluster analyses (single linkage, average linkage, median, centroid or furthest neighbour), or to construct a minimum spanning tree (Gower and Ross 1969), or to print nearest neighbours. Non-hierarchical clustering is available via an exchange algorithm that starts with an initial configuration and then transfers units between classes in order to maximize one of five available criteria.

4.3 *Multivariate analysis*

Directives are available for canonical variate analysis, factor rotation, principal component analysis, principal coordinate analysis and Procrustes rotation. Other techniques, for example canonical correlation analysis and multivariate analysis of variance, are provided by procedures using the general vector and matrix manipulation facilities, which include the calculation of latent roots and vectors of a symmetric matrix and the singular-value decomposition of a rectangular matrix.

4.4 *Non-linear models*

Models for non-linear regression can be fitted by maximum likelihood using one of two derivative-free optimization algorithms. There is also a new directive in Genstat 5, based on methodology from MLP, for fitting a range of standard curves, such as exponential and logistic growth curves.

4.5 *Time series analysis*

Time series may be analysed using the range of ARIMA and seasonal ARIMA models defined by Box and Jenkins. The relationship between series may be investigated by transfer-function models, relating one output series to a number of input series. Special functions and commands are provided to help in model selection and model checking, as well as the estimation of parameters and forecasting from a model. An algorithm for calculating Fourier transforms is available, allowing spectral analysis to be performed.

5. Conclusion

For the use of generalized linear models to achieve widespread acceptance in routine statistical analysis, facilities for GLMs need to be associated with a wide and extensible range of statistical techniques. If the analyst needs to learn a new syntax to do a GLM analysis, there will be a strong temptation to do an approximate - or inappropriate - analysis instead. Genstat has long offered such a range of facilities for batch work and the new version 5 provides convenient interactive use.

A prototype version of Genstat 5 will be demonstrated at the Genstat conference during 23-26 September 1985, at the University of York. Further details about facilities and progress can be obtained from the Genstat Coordinator, Numerical Algorithms Group, Mayfield House, 256 Banbury Road, Oxford, OX2 7DE, United Kingdom.

References

Gower, J.C. and Ross, G.J.S. (1969) Minimum spanning trees and single linkage cluster analysis. *Applied Statistics,* **18**, 54-64.

Lamacraft, R.R. and Payne, R.W. (1980). A new look at data structures for statistical languages. In *Compstat 1980: Proceedings in Computational Statistics* (ed. Marjorie M. Barritt and D. Wishart), 463-469. Vienna: Physica-Verlag.

Lane, P.W. and Nelder, J.A. (1982) Analysis of Covariance and Standardization as Instances of Prediction. *Biometrics,* **38**, 613-621.

Payne, R.W. and Wilkinson, G.N. (1977) A General Algorithm for Analysis of Variance. *Applied Statistics,* **26**, 251-260.

Ross, G.J.S. (1980) *Maximum Likelihood Program.* Harpenden: Rothamsted Experimental Station.

Wilkinson, G.N. (1970). A general recursive procedure for analysis of variance. *Biometrika,* **57**, 19-46.

STATISTICAL MODELLING OF DATA FROM HIERARCHICAL STRUCTURES USING VARIANCE COMPONENT ANALYSIS.

N. T. LONGFORD
CENTRE FOR APPLIED STATISTICS, LANCASTER UNIVERSITY

ABSTRACT

A general statistical modelling framework for variance component analysis of clustered observations (subjects within groups) is set up and demonstrated on a data set originating from a survey of house prices. It may be possible to interface the software used for the data analysis with the new version of GLIM through the $PASS command.

Keywords: variance components, statistical modelling, maximum likelihood.

1. INTRODUCTION.

Survey data frequently involve hierarchical structure owing to either design (stratified sampling) or technicalities of data collection (allocation of subjects to interviewers), or the structure is intrinsically present and relevent to the inference problems for which the data have been collected (e.g. administrative units in nationwide surveys involving educational, health-related, economic, or other issues).

Observations within clusters on any stratum tend to be more homogeneous than between clusters and ordinary least squares (OLS) regression used with such data can lead to substantially biased estimates of regression slopes and their standard errors. The need for variance component modelling of such data has long been recognized but lack of suitable computational algorithms and software has led to developments of various approximate methods (e. g. estimation of the 'design effect'), frequently with negative results.

Following Dempster, Laird, and Rubin (1977) E-M algorithm for (co-)variance component analysis has been developed by Dempster, Rubin, and Tsutakawa (1981), Laird, and Ware (1982), Mason, Wong, and Entwisle (1983), and others. The E-M algorithm tends to converge very slowly, but offers maximum likelihood estimates under very general assumptions.

More recently Goldstein (1985) and Longford (1985) have designed computational algorithms for variance component analysis of hierarchically structured data, which converge rapidly and require only a moderate amount of computation in each iteration. This enables routine use of these procedures for model selection (refitting) even with large data sets. These advantages are demonstrated on a data set using an interactive program/package written by the author.

Model selection can be made on every level of the design, and the variance components involved can be of interest for assessing variability between clusters of each level of the hierarchy and to test whether a sampling level is relevent at all. The conditional means of the random effects associated with the clusters of every level are of interest, and can be used for diagnostic purposes. An application in educational research is discussed by Aitkin and Longford (1985).

2. THE EXAMPLE. HEDONIC HOUSE PRICES.

Harrison and Rubinfeld (1978) have studied the relationship of house prices on quality of the environment (cleanness of air). Their study was based on data from 506 census tracts in the Boston Metropolitan Area, U.S.A. From various census sources they collected observations on 14 variables they believed to be relevent to the problem. Definitions of these variables are given in the Appendix. The appropriateness of using medians (or means) of some of the variables over the census tracts is not discussed here. Harrison and Rubinfeld have used linear regression to explain variation in LMV with the other variables given in the Appendix as explanatory variables.

The data were reanalyzed by Belsley, Kuh, and Welsch (1980) who, after applying various regression diagnostics tools, decided to fit a robust regression model. Belsley et al (1980) have published the data together with a table which indicates that the 506 census tracts belong to 92 towns of various sizes. Some towns contain only 1 census tract; 43 towns have less than 4 census tracts, while 14 towns have 10 or more census tracts. The excessive proportion of large residuals of the OLS regression fit, noted by Belsley et al, is probably caused by extra homogeneity of the values of LMV within towns. This prompted us to reanalyze the data using variance component analysis.

The general variance component model for the design with subjects in groups is set up in Section 3. In Section 4 we give details of our analysis for the Hedonic house price data. For analysis of more complicated hierarchical (multilevel) designs we refer to Longford (1985) and Goldstein (1985).

3. VARIANCE COMPONENT MODELLING (SUBJECTS WITHIN GROUPS).

The modelling framework for variance component analysis is similar to the classical linear regression model

$$y_{ij} = \sum_{k=0}^{P} x_{ij,k}\beta_k + \epsilon_{ij}, \tag{1}$$

where y_{ij} is the value of the y-variate for the subject (census tract) $j = 1, 2, \ldots, n_i$ in group (town) $i = 1, 2, \ldots, N_2 = 92$, $x_{ij,0} \equiv 1$, and $x_{ij,k}$ $(k = 1, 2, \ldots, P)$ represent the explanatory variables; β_k are the regression parameters (intercept β_0 and slopes $\beta_1, \ldots, \beta_P$), and (ϵ_{ij}) is a random sample from $N(0, \sigma_e^2)$. The model (1) takes no account of the clustering of subjects into groups, and it cannot allow for extra homogeneity within groups. The variance component model allows the parameters β_k to vary from group to group; it assumes that

$$y_{ij} = \sum_{k=0}^{P} x_{ij,k}\, \beta_{i,k} + \epsilon_{ij}, \tag{2}$$

where $\underline{\beta}_i = (\beta_{i,0}, \ldots, \beta_{i,P})$, $i = 1, \ldots, N_2$, is a random sample from a (P+1)-variate normal distribution with mean $\underline{\beta} = (\beta_0, \ldots, \beta_P)$ and variance matrix Θ. This random sample is assumed to be independent of (ϵ_{ij}). Any of the random parameters $\beta_{\cdot,k}$ can be 'restricted´ to a constant by specifying their variances as equal to 0. Also, in order to restrict the number of covariance parameters in θ, we will assume that all the random slopes $\beta_{.,k}$ (k = 1, ..., P) are mutually independent, but can be correlated with the random intercept $\beta_{.,0}$. Thus we have

$$\theta = \begin{pmatrix} \theta_0 & \theta_{01} & \theta_{02} & . & . & . & \theta_{0P} \\ \theta_{01} & \theta_1 & 0 & . & . & . & 0 \\ \theta_{02} & 0 & \theta_2 & . & & & . \\ . & . & . & . & . & & . \\ . & . & & . & . & . & . \\ . & . & & & . & . & 0 \\ \theta_{0P} & 0 & . & & . & 0 & \theta_P \end{pmatrix} \qquad (3)$$

This is the minimal parametrization that guarantees invariance with respect to linear transformation of the variables in the model, see Longford (1985).

Thus the model specification consists of the list of variables to be included in the model (the <u>fixed part</u>), and a sublist of these variables for which the corresponding parameters are variable from group to group (the <u>random part</u>) with unknown non-negative variance. The results of a fit are:

1. m. l. estimates of the fixed-effect parameters $\underline{\beta}$,
2. m. l. e. of the subject level variance σ_e^2,
3. m. l. e.´s of the group-level variances θ_0, θ_1, ..., θ_P, and covariances θ_{01}, ..., θ_{0P},
4. value of the likelihood (or deviance = -2*log-likelihood),
5. conditional means of the random effects, i. e.

$$b_{i,k} = E\,(\beta_{i,k} - \beta_k \mid \underline{y}, X; \underline{\beta}, \sigma_e^2, \theta), \qquad (4)$$

for further details see Longford (1985). Also standard errors of the estimates 1. - 3. are obtained (for variances in 2. and 3. we have the standard errors for the standard deviations σ_e, $\sqrt{\theta_0}$, ..., $\sqrt{\theta_P}$). The residual effect attributable to the group i can now be represented by the linear combination

$$\sum_{k=0}^{p} x_{\cdot,k}\, b_{i,k}, \qquad (5)$$

which can form a basis for comparison of groups for a set of explanatory variables $x_{\cdot,k}$ (only the variables k included in the random part of the model are relevent here). This comparison is independent of the values of $x_{\cdot,k}$ if all the slope-variances θ_1, θ_2, ..., θ_p vanish.

4. ANALYSIS OF THE HOUSE PRICES DATA.

Of the 13 explanatory variables (see Appendix) eight are observed on town level (their values are constant within towns). For such a variable the corresponding within-town random regression slope cannot be identified, and so we assume that the variances of these slopes vanish. Thus only 5 variables, CRIM, RM, AGE, B, LSTAT, which have values variable within towns, were considered for the random part of the model.

The value of the -2*log-likelihood for the OLS fit is -300.2. If we allow the intercept $\beta_{\cdot,0}$ to be random (and restrict all the slopes to zero variances), the value of the deviance for the m. l. fit is reduced to -472.0. If also the slopes on CRIM, RM, and LSTAT are allowed to be random, we obtain the value of the deviance -589.5. Further inclusion of variables in the random part leads to very small reduction of the deviance. Estimates of the fixed effects $\underline{\beta}$ and their standard errors (together with their OLS counterparts) are given in Table 1, wich also contains the ´design effect´ (ratio of the variance-component and OLS standard errors for $\underline{\beta}$), and estimates of the variance components. The covariances θ_{01}, ..., θ_{0p} are nuisance parameters, and their estimates are not given.

We note that the design effect varies considerably among the parameters, and it is even smaller than 1 for two of them. The t-ratios for some of the fixed-effect parameters of the OLS and variance component fits differ substantially (CHAS and AGE are cases in point).

TABLE 1. OLS and variance component estimates for the House price data.

Variable	O L S		Var. comp.		Design effect	Estimates of var.comp's θ_k	$\sqrt{\theta_k}$	St. error for $\sqrt{\theta_k}$
	Estimate	St. error	Estimate	St. error				
CRIM	-.0119	.0012	-.0050	.0022	1.78	.00017	.0131	.0025
ZN	.0001	.0005	.0003	.0006	1.10		.0	
INDUS	.0002	.0024	.0027	.0036	1.49		.0	
CHAS	.0914	.0332	-.0195	.0274	.82		.0	
NOXSQ	-.0064	.0011	-.0047	.0013	1.12		.0	
RM	.0063	.0013	.0150	.0017	1.27	.00004	.0064	.0015
AGE	.0001	.0005	-.0020	.0005	.87		.0	
DIS	-.191	.0334	-.1789	.0398	1.19		.0	
RAD	.0957	.0191	.0712	.0227	1.19		.0	
TAX	-.0004	.0001	-.00025	.00016	1.30		.0	
PTRATIO	-.0311	.0050	-.0165	.0077	1.55		.0	
B	.364	.103	.4816	.1715	1.66		.0	
LSTAT	-.371	.025	-.2171	.0314	1.26	.02396	.1548	.0185
INTERCEPT	9.758		9.433			.01359	.1166	.0084

Estimate of subject level variance σ_e^2: .0325 (OLS) , .0116 for variance component analysis

Nearly identical results are obtained if the two variables ZN and INDUS, which have very low t-ratios in all the models fitted, are excluded from the fixed part, or if the variables AGE and/or B are included in the random part of the model.

REFERENCES

Aitkin, M., and Longford, N.T. (1985) Statistical modelling issues in school effectiveness studies. J. Roy. Stat. Soc., to appear.

Belsley, D., Kuh, E., and Welsch, R.E. (1980) Regression Diagnostics - Identifying Influential Data and Sources of Collinearity. Wiley Series in Probability and Mathematical Statistics.

Dempster, A.P., Laird, N.M., and Rubin, D.B. (1977) Maximum likelihood for incomplete data via the EM algorithm. J. Roy. Stat. Soc. B 39, 1 - 38.

Dempster, A.P., Rubin, D.B., and Tsutakawa, R.K. (1981) Estimation in covariance component models. J. Amer. Stat. Assoc. 76, 341 - 353.

Goldstein, H. (1985) Multilevel mixed linear model analysis using iterative generalized least squares. Biometrika, to appear.

Harrison, D., and Rubinfeld, D.L. (1978) Hedonic prices and the demand for clean air. J. Env. Econ. Manag. 5, 81 - 102.

Laird, N.M., and Ware, J.H. (1982) Random-effects models for longitudinal data. Biometrics 38, 963 - 974.

Longford, N.T. (1985) A fast scoring algorithm for maximum likelihood estimation in unbalanced mixed models with nested random effects. Submitted to J. Amer. Stat. Assoc.

Mason, W.M., Wong, G.Y., and Entwisle, B. (1984) The multilevel linear model: A better way to do contextual analysis. Sociological Methodology, Jossey Press, London.

APPENDIX

Definition of model variables (reproduced from Belsley et al 1980, with permission of J. Wiley & Sons).

SYMBOL	DEFINITION
LMV	logarithm of the median value of owner-occupied homes (the y-variate)
CRIM	per capita crime rate
ZN	proportion of town's residential land zoned for lots greater than 25,000 square feet
INDUS	proportion of non-retail business acres per town
CHAS	Charles River dummy variable with value 1 if town bounds on the Charles River
NOXSQ	nitrogen oxide concentration (parts per 10^8)
RM	average number of rooms squared
AGE	proportion of owner-occupied units built prior to 1940
DIS	logarithm of the weighted distances of five employment centres in the Boston region
RAD	logarithm of index of accessibility to radial highways
TAX	full-value property-tax rate (per $10,000)
PTRATIO	pupil-teacher ratio by town
B	$(Bk-0.63)^2$, where Bk is the proportion of blacks in the population
LSTAT	logarithm of the proportion of the population that is of lower status

QUASI-LIKELIHOOD and GLIM

By J.A. NELDER

Imperial College, London, UK

SUMMARY

Quasi-likelihood allows GLMs to be specified by use of the link function and variance function only, or equivalently by specifying the first two moments of the error distribution. Wedderburn's original definition is extended to allow the comparison of different variance functions, and several uses of quasi-likelihood in extending the range of GLMs are described.

Keywords: Quasi-likelihood, deviance, variance function, linear predictor, link function.

1. Introduction

It is an interesting property of the algorithm used in fitting GLMs in GLIM that the distributional assumption of the errors enters only through the variance function of that distribution. Thus given the assumption of Poisson errors for counts, the algorithm uses only the fact that $V(\mu) = \mu$ for the Poisson distribution, and no other characteristics of the distribution appear to be needed. Wedderburn (1974) used this fact to propose a quasi-likelihood Q, defined by

$$Q = \int \frac{y-\mu}{V(\mu)} d\mu \tag{1.1}$$

He showed that estimates obtained by maximizing Q (MQL estimates) have many of the properties of ML estimates. When there exists a distribution of the kind of exponential family that underlies GLMs with a variance function $V(\mu)$, then Q is identical to the likelihood proper. However Q is defined when we have specified only the first two moments of the errors in the form

$$E(y) = \mu$$
$$\mathrm{var}(y) = \phi V(\mu)$$

where ϕ is the dispersion parameter. Gauss made the same step when he replaced an original assumption of full Normality for least squares with one that specified only a constant variance ($V(\mu) = 1$). An obvious use for quasi-likelihood is with over-dispersed data in the form of counts where we assume

$$\mathrm{var}(y) = \phi\mu, \quad \phi > 1$$

The fitting procedure remains the same as for log-linear models, except that the mean deviance is used to estimate ϕ, and asymptotic co-variances are adjusted accordingly. Similarly the use of the heterogeneity factor of probit analysis can be interpreted via a quasi-likelihood for an over-dispersed 'binomial' distribution.

2. Extended Quasi-Likelihood

A deviance can be obtained from a quasi-likelihood in the form

$$D = 2\sum \int_{\mu}^{y} \frac{y-\mu}{V(\mu)}\, d\mu \tag{2.1}$$

(with summation over the observations),
corresponding exactly to the formula obtained from a likelihood proper. Such deviances can be used to compare the fits of models involving different covariates in the linear predictor, and also different link functions. They cannot be used as they stand to compare different <u>variance</u> functions on the same data. For this an extended version of (1.1) is needed, and the form was given by Nelder and Pregibon (McCullagh and Nelder (1983), p. 212) as

$$Q' = \frac{Q - Q_0}{\phi} - \frac{1}{2}\ln(2\pi\phi V(y)) \tag{2.2}$$

Where Q is the Wedderburn QL of (1.1), Q_0 is Q evaluated at $\mu = y$, ϕ is the dispersion parameter, and $V(y)$ is the variance function applied to the observations y. If we multiply (2.2) through by -2 and use (2.1), we can define an <u>extended deviance</u> D' as

$$D' = D/\phi + \sum \ln(2\pi\phi V(y)) \tag{2.3}$$

The use of D' allows us to compare different variance functions on the same data. The expression in (2.2) is in fact the saddle-point approximation to the exponential distributions used in GLMs, whenever such a distribution exists (Daniels (1954)).

It follows that (2.2) is exactly the Normal and inverse Gaussian likelihood for $V(\mu) = 1$ and μ^3 respectively, differs from that of the gamma distribution when $V(\mu) = \mu^2$ by a term depending on ϕ only, and can be obtained for the discrete distributions (Poisson, binomial, negative binomial) by replacing all factorials by their Stirling approximation,

$$k! \approx \sqrt{2\pi k}\,.\,k^k e^{-k} \tag{2.4}$$

We can improve this approximation, and also allow D' to be defined for zero counts (for which ln V(y) would otherwise be infinite) by using

$$k! \approx \sqrt{2\pi(k + 1/6)}\,.\,k^k e^{-k} \tag{2.5}$$

in (2.2) and (2.3). This is equivalent to replacing V(y) by V(y + 1/6) for the quasi-Poisson distribution, with similar adjustments for the binomial and negative binomial distributions (see McCullagh and Nelder (1983), p.214 for details).

The MQL estimates of β in a GLM obtained from Q' are identical to those from Q, because Q' is a linear function of Q with coefficients independent of μ. Further ϕ has MQL estimate $\hat{\phi} = D/N$, where N is the number of observations. By analogy with the Normal case, it might be better to use $\hat{\phi} = D/\nu$, where ν are the d.f. for the deviance. In testing the appropriateness of a particular variance function, it is often useful to embed it in a parameterized family, e.g. $V_\theta(\mu) = \mu^\theta$; testing a prior valuemay then be done by comparing the values of D' for $\theta = \theta_0$ and $\theta = \hat{\theta}$, the MQL estimator.

3. Useful Variance Functions for QLs

For counts the following three forms of variance are of interest:

$$\phi\mu \tag{3.1}$$

$$\mu + \theta\mu^2 \tag{3.2}$$

$$\phi\mu(1 + \theta\mu) \tag{3.3}$$

Of these (3.1) is the over-dispersed 'Poisson' QL, while (3.2) is the variance function of a GLM distribution (for θ known), namely the negative binomial; (3.3) combines the other two, and was used by Bartlett in 1936 for the analysis of insect counts from field trials. In general θ will require to be estimated, and the minimization of D' may be used for this purpose.

For proportions we can similarly extend the basic binomial by replacing

μ in the above expressions by pq, so that (3.1) gives us the binomial with heterogeneity factor etc. Note that Wedderburn's original example used $\phi p^2 q^2$ for the variance, a special case of (3.2) when the second term is dominant.

For continuous measurements the family $V(\mu) = \mu^\theta$ is likely to be of most interest. For θ = 0,2,3 we have the Normal, gamma, and inverse Gaussian distributions respectively and recent work by Tweedie (1981) deals with fractional values of θ.

Multinomial models have been extended, e.g. by Goodhardt, Ehrenberg, and Chatfield (1984), by assuming a Dirichlet distribution for the multinomial parameters. It is easily shown that this has a QL analogue with a variance function identical to that of the multinomial, but with an additional dispersion parameter which is here a function of the multinomial sample sizes.

While the quasi-likelihood formulation is convenient for variance functions which have no exact distributions in the GLM family, uses of QL are potentially much wider than this, and in the remainder of this paper we sketch some futher applications.

4. An Extra Random Component in the Linear Predictor

It is often necessary to consider a linear predictor in a GLM which itself contains a further random element, corresponding perhaps to a latent factor described in distributional terms. Thus suppose we have

$$\eta' = \eta + \varepsilon \quad , \quad \eta = \Sigma \beta_j x_j$$

where ε is a random variable with variance θ, and the corresponding mean μ' is given by $\mu' = g(\eta')$, where g( ) is the inverse link function. Then, approximately

$$E(y) = E[g\,(\eta + \varepsilon)]$$

$$\approx E[g(\eta) + g'(\eta)\varepsilon] = g(\eta)$$

and $\mathrm{var}(y) \approx \mathrm{var}_1(y) + (g')^2.\theta$

where $\mathrm{var}_1(y)$ is the variance of y given μ'

$= \phi\, V(\mu) + (g')^2.\theta$ for a GLM .

Now for a canonical link g' = V, so that

$$\text{var}(y) \approx \phi V(\mu) + \theta V^2(\mu)$$

Thus we can fit such models by replacing the variance function of the 'basic' distribution of errors by one involving an extra squared term in the variance function. Thus for log-linear models with Poisson error we replace $V(\mu) = \mu$ by $\mu + \theta\mu^2$. It follows that the Wedderburn quasi-likelihood here is identical to the likelihood of the negative-binomial distribution, though we have not used a derivation depending on more than the first two moments.

Example: In the proceedings of the last GLIM conference Hinde (1982) analysed some data on fabric faults using a compound model based on giving the Poisson mean a Normal distribution. The computation for this model was quite intensive, involving the use of Gaussian quadrature to evaluate the various integrals involved. By contrast the QL method, using $V = \mu + \theta\mu^2$, is straightforward in GLIM, requiring only the estimation of θ. The resulting fits from the two methods are shown in Table 1. Clearly the two models give very similar results.

Table 1

Comparison on two models for data on fabric faults

Poisson-Normal			QL with $V = \mu + \sigma^2\mu^2$		
b_0	-3.31	1.53	b_0	-3.86	1.44
b_1	0.85	0.24	b_1	0.94	0.23
$\hat{\sigma}$	0.36		$\hat{\sigma}$	0.36	

5. GLMs with More Than One Error Term

Classical Normal models allow extensions to more complex covariance structures than the $\sigma^2 I$ of ordinary regression. In particular designed experiments give rise to models with several error terms, often associated with particular classifications of the experimenal units. The split-plot design, with its two components of error, is a simple example. However, with distributions other than the Normal, there are in general no analogous extensions. Here again QL arguments can come to the rescue.

Consider data in the form of counts that come from a survay with some form of grouping in the sampling. The use of the standard log-linear model ignores possible group differences and their effects on estimates and on their (co)variances. A QL model which treats the group effects as an extra random component can be developed as follows. We add random components to the linear predictor which are identical within groups and independent between groups; this gives a variance matrix for each group of the form

$$V = \text{diag}\,\mu + \theta\mu\mu'$$

This has an inverse of the form

$$V^{-} = \text{diag}\,\mu^{-1} - \frac{\theta}{1+\theta M} 1\,1'$$

where 1 is the vector of ones and $M = \Sigma\mu_i$

The weight matrix for one cluster is given by

$$W = \text{diag}\,\mu\, V^{-}\, \text{diag}\,\mu = \text{diag}\,\mu - \frac{\theta}{1+\theta M}\mu\mu'$$

and hence the contribution to the SSP matrix is

$$X'WX = X'\,\text{diag}\,\mu\; X - \frac{\theta}{1+\theta M} X'\mu\mu' X$$

$$= (\text{SSP within groups}) + \frac{1}{1+\theta M}(\text{SSP between groups})$$

The extra variance component can be chosen to make the mean deviance equal to unity, and both estimates and their (co)variances will be correctly weighted for the group effects.

6. Compound Distributions for Discrete Data

Suppose $u = z_1 + z_2 + \ldots + z_N$

Where the z_i are i.i.d. variables with mean μ_z and variance V_z, while N itself has a distribution with mean μ_N and variance V_N. Then we have

$$E(u) = \mu_N \mu_z$$

$$\mathrm{var}(u) = \mu_Z^2 V_N + \mu_N V_Z$$

Now if the distribution of N has a variance of the form $\phi\mu(1+\theta\mu)$ then so does that of u for

$$\mathrm{var}(u) = \mu_Z^2 \phi\mu_N(1+\theta\mu_N) + \mu_N V_Z$$

$$= \phi'\mu_u(1+\theta'\mu_u)$$

where $\phi' = \phi\mu_Z + V_Z/\mu_Z$ and $\theta' = \theta\phi/\phi'$

Thus the Bartlett variance function is reproduced under this form of compounding, and QL methods may be used to fit the corresponding models. An application is to migration data where N is the number of households moving, z the number of persons per household and u the total number of persons moving.

7. Other Applications

In his review article Pregibon (1984) suggests a major extension of GLMs in which both the mean and the dispersion parameter are modelled in the GLM manner, each with its own link function and linear predictor. It is easy to show that a QL specification of these models leads to a see-saw type of estimation procedure with the means μ_i being modelled given the dispersions ϕ_i and the ϕ_i being modelled using the deviance components from the other fit as y-variate with gamma QL.

I predict that we shall see developments where the so-called Empirical Bayes procedures which lead to the shrinkage estimators for Normal models will be generalized to other GLM distributions using QL formulations. There may also be developments in the domain of time-series for counts, where QL functions based on conditional means and variances will form the basis of fitting procedures. Much remains to be done to develop adequate theory for QL models, but clearly the basic idea is one with many useful applications.

References

Bartlett, M.S. (1936) Some notes on insecticide tests in the laboratory and in the field. J.R. Statist. Soc., **3**, Suppl. 185-194.

Goodhardt, G.J., Ehrenberg, A.S.C. and Chatfield, C. (1984) The Dirichlet: a comprehensive model of buying behaviour. J.R. Statist. Soc.(A), **147**, 621-643

Daniels, H.E. (1954) Saddle-point approximations in Statistics. Ann. Math. Statist., **25**, 631-650.

Hinde, J. (1982) Compound Poisson regression models. GLIM 82 Proceedings, 109-121. Springer-Verlag, New York.

McCullagh, P. and Nelder, J.A. (1983) Generalized Linear Models. Chapman and Hall, London.

Tweedie, M.C.K. (1981) An index which distinguishes between some important exponential families. Proc. Ind. Stat. Inst. International Conference, 580-604.

Wedderburn, R.W.M. (1974) Quasi-likelihood functions, generalized linear models and The Gauss-Newton method. Biometrika, **61**, 439-447.

GLIM for Latent Class Analysis

By JUNI PALMGREN and ANDERS EKHOLM*

National Public Health Institute, Finland

Univ. of Helsinki, Finland

SUMMARY

We show that latent class models are exponential family nonlinear models. These are extended generalized linear models with the link function substituted by an observationwise defined nonlinear function of the model parameters. Latent class models can be fitted using the OWN-facility in GLIM. We analyse a set of data which Clogg and Goodman (1984) fitted by an EM algorithm. The necessary GLIM macros are discussed.

Keywords: Exponential family nonlinear model; Extended generalized linear model; GLIM; IRLS; Latent class analysis

1. Introduction

Consider the data in Table 1, first presented by Solomon (1961). They summarize the answers of 2 982 New Jersey high-school seniors in 1957 to the following four dichotomous items:

A: The development of new ideas is the scientist's greatest source of satisfaction.

B: Scientists and engineers should be eliminated ~~from the military draft.~~

C: The scientist will make his maximum contribution to society when he has freedom to work on problems which interest him.

D: The monetary compensation of a Nobel Prize winner in physics should be at least equal to that given popular entertainers.

The code numbers 1 and 2 denote respectively "agree" and "disagree". The respondents are divided into the upper and the lower half by an IQ vocabulary test.

Clogg and Goodman (1984) have recently analysed Table 1 as an example of fitting "simultaneous latent structure models". The essence of their models can be expressed in one equation using the following notation. Let A, B, C, D denote the four binary responses, G the membership in the upper (coded 1) or lower (coded 2) IQ half, and Z a latent attitude to scientific freedom, with levels "for scientific freedom" (coded

*Present address: Department of Statistics, University of Helsinki, Aleksi 7, SF-00100 Helsinki, Finland

TABLE 1

Cross classification of high-school seniors according to four dichotomised response items, for the upper and the lower IQ half

Item response				Group	
A	B	C	D	High IQ	Low IQ
1	1	1	1	122	62
1	1	1	2	68	70
1	1	2	1	33	31
1	1	2	2	25	41
1	2	1	1	329	283
1	2	1	2	247	253
1	2	2	1	172	200
1	2	2	2	217	305
2	1	1	1	20	40
2	1	1	2	10	11
2	1	2	1	11	11
2	1	2	2	9	14
2	2	1	1	56	31
2	2	1	2	55	46
2	2	2	1	64	37
2	2	2	2	53	82
Total				1 491	1 491

1), and "against scientific freedom" (coded 2). The generic symbol for the levels of A is a, and analogously for the other variables.

The basic equation is for a, b, c, d, g, z = 1, 2

$$P(A=a,\ B=b,\ C=c,\ D=d \mid G=g) = \Sigma_z P(Z=z \mid G=g) \times$$

$$P(A=a \mid G=g\ Z=z) P(B=b \mid G=g\ Z=z) P(C=c \mid G=g\ Z=Z) P(D=d \mid G=g\ Z=z). \qquad (1.1)$$

There are 2 x 16 = 32 different probabilities on the left hand side, and the right hand side is made up of 2 + 2 x 2 x 4 = 18 different probabilities. Since the number of respondents in each IQ half is fixed to 1 491 equation (1.1) implies a model for 30 independent observations, with 18 parameters and 12 degrees of freedom. Clogg and Goodman develop and use an EM algorithm for fitting this model. Subsequently they fit models with restrictions on the parameters.

The point of the present paper is that latent class models of the kind in equation (1.1) bear a close resemblance to generalized linear models. Thompson and Baker (1981), Jörgensen (1984a), Cox (1984) and Green (1984) have shown that most of the theory developed for generalized linear models carries over to - in the terminology of Palmgren and Ekholm (1984) - exponential family nonlinear models. An iteratively reweighted least squares algorithm can be used for fitting them. The computations can be done in GLIM, using the OWN-facility, thereby giving the user access to the output facilities for e.g. inspection of covariances of estimates, and analysis of residuals.

In Section 2 we review briefly the set up of exponential family nonlinear models and relate them to latent class models. Section 3 indicates the estimation theory and the construction of the necessary macros. In Section 4 we return to the example introduced above, and the Appendix contains the GLIM-program with comments. Section 5 discusses briefly some closely related approaches.

2. Exponential Family Nonlinear Models

We assume that the reader is familiar with the theory of generalized linear models in the sense of McCullagh and Nelder (1983, Ch. 2). We write Y_i for the i:th random count with $i = 1,\ldots,n$. In the example above $n = 32$. The random counts are assumed to be independent Poisson variates or to follow a multinomial or a product multinomial distribution. More generally, they might follow any discrete one-parameter exponential family distribution with known or unknown scale factor.

Let $\underline{x}_i = (x_{i1},\ldots,x_{ip})'$ be a vector of known and fixed values of p ($< n$) explanatory variables coupled to the i:th count. Also, let $\underline{\beta} = (\beta_1,\ldots,\beta_p)'$ be a vector of model parameters. Finally write $\mu_i = E(Y_i)$ for $i = 1,\ldots,n$.

A generalized linear model connects μ_i to $\underline{x}_i$ and $\underline{\beta}$ by

$$g(\mu_i) = \underline{x}_i'\underline{\beta}, \qquad (2.1)$$

where the link function g( ) is monotonic and differentiable. Now, consider the latent class model (1.1). The left hand side is clearly $\mu_i/1491$ where i runs through the 32 possibilities given by (a,b,c,d,g). The vector $\underline{\beta}$ consists of the 18 probabilities on the right hand side. The vector $\underline{x}_i$ indicates to which IQ group the i:th count refers. The right hand side is accordingly a sum of products of the model parameters. It is, clearly, impossible to find a link function that reduces a sum of products to an expression linear in the model parameters.

Instead, we work with observationwise defined nonlinear functions of the model parameters. Generally, we substitute (2.1) by

$$\mu_i = h_i(\underline{x}_i,\underline{\beta}), \qquad (2.2)$$

where the functions $h_i(\ )$, for $i = 1,\ldots,n$ are differentiable with respect to all the β's. In the example the $h_i(\ )$ functions are given by the right hand side of equation (1.1).

A general formulation of latent class models relating them to equation (2.2) is as follows. Let $Np_i(\underline{x}_i;\underline{\beta})$ be the expected value of the i:th random count, with

$\Sigma_i p_i(\underline{x}_i;\underline{\beta}) = 1$, and N the fixed number of categorized objects. The random counts are arranged in a multiway table. The latent structure is introduced by assuming that there is one or several further unobservable dimensions with a total of m classes.

Let $\pi_k(\underline{x})$ for $k = 1,\ldots,m$ be the probability that an object belongs to the latent class k for any fixed $\underline{x}$. The conditional probability that an observation belongs to category i of the observed table given that it belongs to the latent class k is denoted $\pi_{i|k}(\underline{x}_i)$. We set $\Sigma_k \pi_k(\underline{x}) = \Sigma_i \pi_{i|k}(\underline{x}_i) = 1$.

To specify identifiable models in this way a structure has to be put either on $\pi_k(\underline{x})$ or on $\pi_{i|k}(\underline{x}_i)$ or on both. The ubiquitous assumption is that the manifest dimensions are independent given the latent class. This assumption is exemplified in equation (1.1). It is commonly referred to as local independence. Further structure on $\pi_k(\underline{x})$ and $\pi_{i|k}(\underline{x}_i)$ can be introduced by expressing them both as functions of a smaller set of model parameters $\underline{\beta}$ as $\pi_k(\underline{x};\underline{\beta})$ and $\pi_{i|k}(\underline{x}_i;\underline{\beta})$ respectively.

The functions $h_i(\;)$ in equation (2.2) are of the following form for latent class models

$$\mu_i = N\,\Sigma_k \pi_k(\underline{x}_i;\underline{\beta})\pi_{i|k}(\underline{x}_i;\underline{\beta}). \tag{2.3}$$

3. Estimation and GLIM Macros

We define for $i = 1,\ldots,n$ and $v = 1,\ldots,p$

$$h_{iv} = \partial\mu_i/\partial\beta_v = \partial h_i(\underline{x}_i;\underline{\beta})/\partial\beta_v \tag{3.1}$$

as the elements of an n x p local model matrix H. These elements are, in general, functions of the unknown vector parameter $\underline{\beta}$. We assume that the model matrix H is of full rank throughout the parameter space, cf. Clogg and Goodman (1984, p. 766).

Green (1984), Jörgensen (1984b) and Palmgren and Ekholm (1984) have shown that maximum likelihood estimates of $\underline{\beta}$ can be obtained by iteratively solving the following linear equations for $t = 0,1,\ldots$

$$(H_t' V_t H_t)\underline{\beta}_{t+1} = H_t' V_t \underline{z}_t,$$

where $\underline{z}_t$ is a vector of local dependent variates and V_t is a diagonal matrix of local weights. The three local arrays H_t, $\underline{z}_t$ and V_t are all evaluated at the current parameter values $\underline{\beta}_t$.

Writing $\underline{y}$ and $\underline{\mu}$ respectively for the vectors of observed and fitted values of the random counts the vector $\underline{z}_t$ is given by

$$\underline{z}_t = H_t\underline{\beta}_t + \underline{y} - \underline{\mu}_t. \tag{3.2}$$

The expression for the i:th element of the weight matrix V depends only upon the distribution of Y_i, and is in the Poisson case $1/\mu_i$. Palmgren and Ekholm (1984, p. 9) discuss in detail a factorization of (3.1) in the case of a simple link. The factorization then makes it possible to use a global model matrix, and accordingly ordinary GLIM specifications.

To fit latent class models we have to use the OWN-facility in GLIM. The macros for the variance function and for the deviance increment (Baker and Nelder, 1978, Sec. 18.2) are specified exactly as for ordinary Poisson models. The derivative of the inverse of the link function is set to 1. This follows from the way the local dependent variate is defined in (3.2), and the local weights are defined beneath (3.2).

The macro for the fitted values should give these as functions of the current parameter estimates using equation (2.3). Also, this macro must contain the updating of the local model matrix using equation (3.1). The details of this exercise are described in the Appendix.

4. Numerical example

First we fitted model (1.1) to the data in Table 1 using the method described in Section 3. This gave us the same results as Clogg and Goodman (1984, p. 768) report. The deviance of model (1.1) is 20.32 on 12 degrees of freedom.

Next, we noticed that items B and D deal with the gratification that the society should provide scientists, while items A and C concern their inner gratification. We therefore restricted the probability of a positive response to items A and C to be the same in the positive latent class in both IQ groups. The unrestricted model indicated that the only further reasonable restriction would be to set the probability of a positive response to item A in the negative latent class equal in the two IQ groups. We thus, for illustrative purposes, have a model with both within group and between group restrictions.

The results for this model are reported in Table 2. The deviance is 22.35 on 16 degrees of freedom. The fit is of similar order as for the model with two latent classes which Clogg and Goodman report. All standardized residuals are below 1.8 in absolute value.

Note that a positive answer to item B has very low odds in both latent classes in both groups. The conclusion might be that item B taps the attitude to the military draft rather than to scientific freedom.

Further, there are eight correlations between the parameter estimates that are above 0.8 in absolute value. Five of these involve either of the probabilities for the latent classes.

The GLIM program for this model is given in the Appendix.

TABLE 2

Estimated parameters for model (1.1) with restrictions

Standard errors in parentheses.

Group	Latent class	P(Z=z|G)	P(A=1|G,Z)	P(B=1|G,Z)	P(C=1|G,Z)	P(D=1|G,Z)
High IQ	"For"	0.40 (0.09)	0.88* (0.02)	0.34 (0.05)	0.88* (0.02)	0.68 (0.05)
	"Against"	0.60 (0.09)	0.77** (0.01)	0.10 (0.02)	0.42 (0.06)	0.45 (0.03)
Low IQ	"For"	0.51 (0.12)	0.88* (0.02)	0.21 (0.02)	0.88* (0.02)	0.53 (0.04)
	"Against"	0.49 (0.12)	0.77** (0.01)	0.13 (0.02)	0.14 (0.16)	0.36 (0.02)

The parameters with stars are restricted to have the same values

Clogg and Goodman (1984, p. 769) treat another example in the same paper. The models they fit to those data are, however, not genuine latent class models, and can be fitted by ordinary GLIM specifications. Palmgren and Ekholm (1984) give three examples, complete with programs, of latent structure models for which one has to use the OWN-specification.

5. Discussion

When Thompson and Baker (1981, p. 125) first extended the theory for generalized linear models by the concept of a composite link, one important impetus seems to have been the notion of a "mixed up" contingency table. Some cells of a mixed up table are indistinguishable so that the expected value of the observed frequency is a straight sum of the expected frequencies in these indistinguishable cells. Roger (1983) simplified the way of analysing mixed up tables in GLIM.

We, however, understand that latent class models cannot be seen as mixed up contingency tables and thus cannot be handled by a linear composite link nor by Roger's modification of it.

A latent class model for a contingency table sees the observed table as a mixture of two or several layers of similar tables. The expected frequencies are not straight sums of the corresponding expected frequencies in the layers but mixtures of them with unknown mixing probabilities.

Burn (1984) pointed out that a certain latent class model is an example of a bilinear composite link, a generalization of the linear composite link mentioned by

Thompson and Baker (1981, p. 129) in connection with models for grouped normal data.

We recognize that model (1.1) could be formulated as a bilinear composite link. Using this notion it would be natural to work with logs of the conditional probabilities for the manifest categories given the latent ones. We have chosen to work with the untransformed conditional probabilities and have not attempted the factorization of the local model matrix (3.1) suggested by the bilinear composite link formulation. For moderate sized problems the programming of the unfactorized local model matrix is a straightforward task. This approach, however, lacks the program generality inherent in Roger's and Burn's work.

Green (1984, p. 155) points out that the choice of parametrization can have a dramatic effect on the convergence of the IRLS algorithm. This question, certainly, deserves consideration for latent class models.

References

Baker, R. J. and Nelder, J. A. (1978) The GLIM System Release 3. Oxford: Numerical Algorithms Group.

Burn, R. (1984) Fitting a logit model to data with classification errors. Glim Newsletter, Issue no **8**, 44-47.

Clogg, C. C. and Goodman, L. A. (1984) Latent structure analysis of a set of multidimensional contingency tables. J. Amer. Statist. Ass. **79**, 762-771.

Cox, C. (1984) Generalized linear models - the missing link. Appl. Statist. **33**, 18-24.

Green, P. J. (1984) Iteratively reweighted least squares for maximum likelihood estimation, and some robust and resistant alternatives. J.R. Statist. Soc. B, **46**, 149-192.

Jörgensen, B. (1984a) Maximum likelihood estimation and large-sample inference for generalized linear and nonlinear regression models. Biometrika **70**, 19-28.

Jörgensen, B. (1984b) The delta algorithm and GLIM. Int. Statist. Rev. **52**, 283-300.

McCullagh, P. and Nelder, J. A. (1983) Generalized Linear Models. London: Chapman and Hall.

Palmgren, J. and Ekholm, A. (1984). Exponential family nonlinear models for categorical data with errors of observation. Univ. of Helsinki, Dept. of Statist., Research Report n:o **52**.

Roger, J. H. (1983) Composite link functions with linear log link and Poisson error. Glim Newsletter, Issue no **7**, 15-21.

Solomon, H. (1961) Classification procedures based on dichotomous response vectors. In Studies in Item Analysis and Prediction (H. Solomon, ed.). Stanford, Calif.: Stanford University Press.

Thompson, R. and Baker, R. J. (1981) Composite link functions in generalized linear models. Appl. Statist. **30**, 125-131.

```
!APPENDIX
    !
    !The GLIM program for model (1.1) with restrictions
    !P(A=1/G=1,Z=1)=P(C=1/G=1,Z=1)=P(A=1/G=2,Z=1)=P(C=1/G=2,Z=1) and
    !P(A=1/G=1,Z=2)=P(A=1/G=2,Z=2) fitted to the data in Table 1.
$UNITS 32
$DATA Y
$READ
122 20 329 56 33 11 172 64 68 10 247 55 25 9 217 53
62 14 283 31 31 11 200 37 70 11 253 46 41 14 305 82
$CAL %N=1491: %I=0
$MACRO M1 $SWI %I MEXT
$VAR 18 P $CAL P(1)=%PE(1):P(2)=P(6)=P(11)=P(15)=%PE(2)
$CAL P(3)=P(12)=%PE(3):P(4)=%PE(4):P(5)=%PE(5):P(7)=%PE(6)
$CAL P(8)=%PE(7):P(9)=%PE(8):P(10)=%PE(9):P(13)=%PE(10)
$CAL P(14)=%PE(11):P(16)=%PE(12):P(17)=%PE(13):P(18)=%PE(14)
    !The vector P has 18 elements corresponding to the probabilities
    !on the right hand side of (1.1). The restrictions above are
    !imposed on P.
$VAR 16 A B C D J
$CAL A=%GL(2,1):B=%GL(2,2):C=%GL(2,4):D=%GL(2,8):J=%GL(16,1)
$VAR 16 XA1 XA2 XB1 XB2 XC1 XC2 XD1 XD2 M $CAL %G=0:%L=0
    !The vectors XA1-XD2 and M are used in macro GROUP to compute %FV
    !and the column vectors in the matrix H defined in (3.1)
$CAL %S=1 $WHILE %S GROUP ! The macro GROUP is invoked once for
    !each group. Thus we avoid writing expressions for %FV and the
    !column vectors of H separately for each group.
$CAL A1=A1+E1:A2=A2+E2 !See comment in macro GROUP for between group
    !restrictions.
$CAL LP1=%PE(1)*P1+%PE(2)*A1+%PE(3)*A2+%PE(4)*B1+%PE(5)*B2
$CAL LP2=%PE(6)*C2+%PE(7)*D1+%PE(8)*D2
$CAL LP3=%PE(9)*P2+%PE(10)*F1+%PE(11)*F2
$CAL LP4=%PE(12)*G2+%PE(13)*H1+%PE(14)*H2
$CAL %LP=LP1+LP2+LP3+LP4 $DEL LP1 LP2 LP3 LP4 $CAL %I=1
    !The linear predictor %LP is the product of the parameter vector
    !%PE and the local model matrix H (cf. 3.2).
$ENDMAC
$MACRO GROUP
    !The scalar %G has value 0 in group 1 and value 9 in group 2.
    !The scalar %L has value 0 in group 1 and value 16 in group 2.
$CAL XA1=(A-1)*(1-P(2+%G))+(2-A)*P(2+%G)
$CAL XA2=(A-1)*(1-P(3+%G))+(2-A)*P(3+%G)
$CAL XB1=(B-1)*(1-P(4+%G))+(2-B)*P(4+%G)
$CAL XB2=(B-1)*(1-P(5+%G))+(2-B)*P(5+%G)
$CAL XC1=(C-1)*(1-P(6+%G))+(2-C)*P(6+%G)
$CAL XC2=(C-1)*(1-P(7+%G))+(2-C)*P(7+%G)
$CAL XD1=(D-1)*(1-P(8+%G))+(2-D)*P(8+%G)
$CAL XD2=(D-1)*(1-P(9+%G))+(2-D)*P(9+%G)
$CAL M=%N*(P(1+%G)*XA1*XB1*XC1*XD1+(1-P(1+%G))*XA2*XB2*XC2*XD2)
$CAL %FV(J+%L)=M $DEL M  $VAR 32 E1 E2
    !The vectors P1,..,H2 below are the columns in the local model
    !matrix H. P1 and P2 associate to the probabilities P(Z=1/G=1)
    !and P(Z=1/G=2). A1 and E1 associate to the probabilities
    !P(A=1/G=1,Z=1) and P(A=1/G=2,Z=1) etc. for A2-D2 in the first
    !group and E2-H2 in the second group. Note, that there are no
    !vectors C1 and G1 due to within group restrictions. The between
    !group restrictions are handled outside macro GROUP by setting
    !A1=A1+E1 and A2=A2+E2
$VAR 16 PP $CAL PP=%N*(XA1*XB1*XC1*XD1-XA2*XB2*XC2*XD2)
$CAL P1(J+%L)=PP*%EQ(%L,0):P2(J+%L)=PP*%EQ(%L,16) $DEL PP
$VAR 16 Q $CAL Q=(A-1)+(C-1)*(5-4*A)+(3-2*A)*(3-2*C)*2*P(2)
```

```
    !The expression Q contains the derivative of XA1*XC1 with re-
    !spect to the parameter P(2). Note, that in this way the within
    !group restriction P(6)=P(2) is taken into account when compu-
    !ting the local model matrix H.
$VAR 16 AA1 $CAL AA1=%N*P(1+%G)*Q*XB1*XD1
$CAL A1(J+%L)=AA1*%EQ(%L,0):E1(J+%L)=AA1*%EQ(%L,16) $DEL AA1
$VAR 16 AA2 $CAL AA2=%N*(1-P(1+%G))*(3-2*A)*XB2*XC2*XD2
$CAL A2(J+%L)=AA2*%EQ(%L,0):E2(J+%L)=AA2*%EQ(%L,16) $DEL AA2
$VAR 16 BB1 $CAL BB1=%N*P(1+%G)*(3-2*B)*XA1*XC1*XD1
$CAL B1(J+%L)=BB1*%EQ(%L,0):F1(J+%L)=BB1*%EQ(%L,16) $DEL BB1
$VAR 16 BB2 $CAL BB2=%N*(1-P(1+%G))*(3-2*B)*XA2*XC2*XD2
$CAL B2(J+%L)=BB2*%EQ(%L,0):F2(J+%L)=BB2*%EQ(%L,16) $DEL BB2
$VAR 16 CC2 $CAL CC2=%N*(1-P(1+%G))*(3-2*C)*XA2*XB2*XD2
$CAL C2(J+%L)=CC2*%EQ(%L,0):G2(J+%L)=CC2*%EQ(%L,16) $DEL CC2
$VAR 16 DD1 $CAL DD1=%N*P(1+%G)*(3-2*D)*XA1*XB1*XC1
$CAL D1(J+%L)=DD1*%EQ(%L,0):H1(J+%L)=DD1*%EQ(%L,16) $DEL DD1
$VAR 16 DD2 $CAL DD2=%N*(1-P(1+%G))*(3-2*D)*XA2*XB2*XC2
$CAL D2(J+%L)=DD2*%EQ(%L,0):H2(J+%L)=DD2*%EQ(%L,16) $DEL DD2
$DEL XA1 XA2 XB1 XB2 XC1 XC2 XD1 XD2 M
$CAL %G=%G+9:%L=%L+16:%S=%EQ(%G,9)-1
$ENDMAC
$MACRO M2 $CAL %DR=1 $ENDMAC
$MACRO M3 $CAL %VA=%FV $ENDMAC
$MACRO M4
$CAL %DI=2*(%YV*%LOG(%YV/%FV)+%FV-%YV)
$ENDMAC
$MACRO MEXT $EXTRACT %PE $ENDMAC
$YVAR Y $OWN M1 M2 M3 M4 $SCA 1
$DATA 14 %PE $READ
0.36 0.90 0.76 0.34 0.12 0.42 0.68 0.46
0.62 0.21 0.11 0.09 0.52 0.33
    !The estimates from the unrestricted model are used as starting
    !values.
$CAL %LP=P1=A1=A2=B1=B2=C2=D1=D2=P2=F1=F2=G2=H1=H2=0
$FIT P1+A1+A2+B1+B2+C2+D1+D2+P2+F1+F2+G2+H1+H2-%GM
    !The FIT directive contains the column vectors in the H matrix.
    !The degrees of freedom reported by GLIM have to be corrected for
    !the number of fixed margins (2).
$DISP E
       SCALED
CYCLE  DEVIANCE      DF
   3     22.35       18

      ESTIMATE      S.E.       PARAMETER
   1  0.3997       .9234E-01   P1
   2  0.8837       .2211E-01   A1
   3  0.7749       .1294E-01   A2
   4  0.3435       .4998E-01   B1
   5  0.1043       .1971E-01   B2
   6  0.4238       .6193E-01   C2
   7  0.6798       .4664E-01   D1
   8  0.4490       .2703E-01   D2
   9  0.5091      0.1181       P2
  10  0.2137       .2489E-01   F1
  11  0.1254       .1528E-01   F2
  12  0.1367      0.1568       G2
  13  0.5299       .3981E-01   H1
  14  0.3645       .2201E-01   H2
  SCALE PARAMETER TAKEN AS     1.000

$RETURN
```

GLIM 3.77

By CLIVE PAYNE and JANET WEBB

Social Studies Faculty Centre, University of Oxford;
NAG Ltd, Oxford

Summary

A new release of the GLIM program, GLIM 3.77, is described. Modifications and extensions to the previous release are given.

Keywords: GLIM, generalised linear models; statistical software.

Introduction

A major new version of GLIM, GLIM 3.77, will be released later this year. The system is written in FORTRAN 77 and will run on a wide range of machines including many microcomputers. A number of changes have been introduced to the program to make GLIM more robust and more efficient. Details of availability and pricing may be obtained from the distributors, NAG Ltd, Oxford.

The GLIM manual has been completely revised to incorporate the modifications and extensions to the current version, 3.12. The new manual will include a detailed and comprehensive Reference Section, a revised version of the 3.12 User Guide which incorporates the changes to 3.77 and an Introductory Guide which introduces new users to GLIM 3.77. This article is an extended version of Appendix E of the new manual which describes the differences between Releases 3.12 and 3.77; it is intended as a preliminary conversion guide for 3.12 users.

1. The Language

1.1 The GLIM character set has been extended. Lower case alphabetic characters are now permitted and are treated as equivalent to the corresponding upper case characters. The character @ is used in fault messages to represent an invalid keystroke input.

1.2 There is now no alternative directive symbol.

1.3 Identifier names may now include the underline character (except as the first character).

1.4 E-format numbers may now be used on input.

1.5 Formal arguments for macros can now be of the form **%scalar** as well as **%digit**; the rounded value of **scalar** must be in the range 1-9. This facility allows a macro to scan through its argument list or to reference a parameter chosen as the macro is running.

1.6 The syntax of two existing commands, PLOT and LOOK, has been extended to include option lists. These give extended control over the actions taken by the directives. Three new commands - TABULATE, TPRINT and HISTOGRAM - also have this syntax. Option lists have the form:

([**options**]s) where **option** is **name=setting**

Example:

```
$histogram (ylim=0,40 cols=5) X
```

Any selection of available options for a directive may appear, in any order.

1.7 The syntax of the GLIM 3.12 directive PRINT and two new directives (TABULATE, TPRINT) now includes phrases. These qualify items and make explicit the function of the item in the directive.

Example:

```
$tabulate the X mean
```

where the phrase 'the mean' qualifies the identifier X.

2. Changes to 3.12 Directives

With the four exceptions marked '+' below, Release 3.77 is upwardly compatible with 3.12.

2.1 ACCURACY. The accuracy setting specified now affects the output from all directives.

2.2 CALCULATE. Additional relational and logical operators are provided to complement the functional forms. The argument of a monadic function may now appear without brackets. Operator precedences have been updated. The new operators are:

>	equivalent to %GT
>=	%GE
<	%LT

<=	equivalent to	%LE
==		%EQ
/=		%NE

Three logical operators have been added and are defined as follows:

& for AND	X&Y is equivalent to	%NE(X,0)*%NE(Y,0)
? for INCLUSIVE OR	X?Y	1-%EQ(X,0)*%EQ(Y,0)
/ for NOT	/A	%EQ(A,0)

2.3 CYCLE and RECYCLE. Tolerances for convergence and the detection of intrinsic aliasing can now be set by the user.

2.4 DATA. There is now no limit (except size of workspace) on the number of identifiers which may be declared.

2.5 DISPLAY. The output has been enhanced and the layout improved. The character '1' is now used in place of '%GM' whenever parameter estimates are output.

2.6 DUMP(+). Only the currently used portions of the data space are stored, resulting in appreciably smaller dump files. The files produced are not compatible with GLIM 3.12, nor are GLIM 3.12 dump files compatible with GLIM 3.77. A dumped program may now be subsequently restored to a program with a different size of workspace, provided the new workspace is large enough to accommodate the dumped data.

2.7 ENVIRONMENT. The output has been improved and three additional options are provided (C,E,G). Option C lists settings for input and output channels and indicates the status of the ECHO, VERIFY, WARN and HELP switches and whether their output is being sent to the TRANSCRIPT file (see 3.13 below). Option E gives a description of the PASS facility (see 3.9). Option G gives information on the graph plotting facilities, if implemented (see 3.2).

2.8 FIT. The output has been improved. The character '1' is now accepted as a substitute for '%GM'. The change in deviance and d.f is now given when the only change in a model is that the structure formula begins with an operator. The effective number of observations is given if this is different from the number of units.

2.9 FORMAT. Some syntax restrictions have been removed and a FORTRAN format statement can now be up to 238 characters long.

2.10 INPUT and REINPUT. The restriction that a subfile identifier must not be the same as that of an existing vector or macro has now been removed.

2.11 LOOK. The user now has more control over the layout of the output produced. The table of vector values can have borders around it and between columns which may be labelled with identifier names. Alternatively, all labelling information can be suppressed.

2.12 PLOT. The user now has more control over the layout of a plot and additional types of plot can be produced. The range of values plotted on both axes can be restricted to give a windowing facility. Plotting symbols can be specified by the user. Overlaid plots can be produced for the distinct values of a factor. Plots are now scaled by default on the basis of non-restricted points (+).

2.13 PRINT. The facilities have been extended to give more control over the formatting of output by the inclusion of phrases in the directive. Margins and word-break can be set. Numbers can be output in specified positions on a line with full control over justification and layout. Text strings and the contents of macros can be output either as a continuous stream or on a line-by-line basis.

2.14 SORT. The second and third arguments may now be any integer number or scalar; these can be used to produce cyclic shifts in the vector being sorted.

2.15 TERMS(+). This directive has been withdrawn.

2.16 USE(+). This directive now has the same syntax as ARGUMENT so that macro arguments can be supplied at the same time as the macro is invoked.

3. New Directives

Those marked with '*' are not available on all computer systems. Items in square brackets in syntax definitions are optional.

3.1 ASSIGN. Assigns a list of values to a vector; the list may be a concatenation of vectors. The syntax is:

$assign **identifier = list of values**

where **list of values** may contain numbers, scalars, variate or factor names, in any combination.

Examples:

```
$assign RANG = 1,100              ! length 2
$assign VALS = 1.5,%B,%C,%D       ! length 4
$assign VALS = VALS,%E,%F         ! length now 6
```

3.2 GRAPH(*). Produces plots on a graphical device. Initially this is available for use with the GINO graphics package only. Options are available to grid a graph, to plot a series of graphs on the same axes and to annotate the axes with titles and scale values. Separate plots (on the same axes) can be generated for points corresponding to each level of a factor.

3.3 GROUP. Recodes a vector into a factor with values assigned to contiguous intervals of the input vector. The syntax is:

$group [**vec2**]=**vec1** [intervals [*] **vec3** [*]][values **vec4**]

Vec1 is the vector to be recoded into the output vector **vec2**. **Vec3** specifies the lower end points of each interval and is introduced by the phrase-name 'intervals'. **Vec4**, introduced by the phrase-name 'values', specifies the integer values 1,2,... to be assigned to the intervals. The optional left/right characters '*' are used to specify lower/upper end-points at $-\infty/+\infty$. There are suitable defaults for omitted items.

Example:

```
$assign X = 5,9,12,35,5
:       INTS = 20,40
:       LEVS = 2,1
$group GX = X intervals * INTS values MIDS
```

forms a factor GX from the vector X with values 2,2,2,1,2 corresponding to the intervals $(-\infty,20]$ and $(20,40]$.

3.4 HISTOGRAM. Produces histograms of vectors (with weights if required) which may be partitioned according to the levels of a factor. The syntax is:

$histogram [**option list**] [**vector**[/**weight**]]s
['**string**'[**factor**]]

The directive plots the joint histogram of the **vector**(s) (weighted by their **weight** vectors, if supplied), using the symbols supplied by **string** partitioned as implied by the levels of the **factor**. The STYLE option determines the layout of the plot which can be enclosed in a box with or without annotation and axes. Options are available to specify lower and upper limits in the histogram (YLIMIT), the smallest and largest frequencies to be plotted (XLIMIT), the maximum number of columns to be used (COLS), the number of rows in the histogram (ROWS) and whether values lying outside the main body of the histogram are also plotted together in extra intervals (TAILS). Similar options are available for the enhanced PLOT directive.

Examples:

```
$histogram (ylimit=0,40 rows=8 tails=1,1 style=1) Y
```

will restrict the plot to Y values between 0 and 40 in 8 rows, but showing both the grouped upper and lower tails as extra rows. The histogram will be fully annotated and enclosed in a box.

```
$histogram Y '123' A
```

plots the histogram of Y for three levels of the factor A.

3.5 LAYOUT(*). Determines the layout of plots on a graphical device. The user can change the dimensions and positions of a graph and choose among the available (conceptual) pens; by scaling a graph to be larger than the plotting area of the output device, an enlargement of a section of the graph can be obtained.

3.6 MANUAL(*). Prints selected sections of the GLIM Manual.

3.7 MAP. Recodes a vector into a variate with values assigned to contiguous intervals of the input vector. This directive has the same syntax as GROUP (3.3); the only difference is that the output vector is defined as a variate so that the values assigned to the intervals are not restricted to integer numbers.

Example:

With directives in (3.3) above

```
$assign MIDS = 10,30
$map MX = X intervals * INTS values MIDS
```

recodes X into a variate MX of length 5 with value 10 for X in the interval $(-\infty, 20]$ and value 30 for the interval $(20,40]$.

3.8 PAGE. A switch to control scrolling of output for interactive use at a terminal. When the number of lines output to the terminal exceeds the page height currently set by $OUTPUT, the program halts and asks for a prompt before continuing. This facility can be switched on or off by the PAGE directive.

3.9 PASS(*). Passes GLIM data structures to and from a user's own FORTRAN program. Thus the user can carry out data manipulation operations or special analyses which cannot be performed easily in GLIM.

3.10 SET. Sets mode of use to either batch or interactive.

3.11 TABULATE. Produces a tabular structure from an input vector (with optional weighting). The structure contains a statistic chosen from the list - mean, total, variance, standard deviation, minimum, median and other percentiles and interpolated values. The structure can be classifed by one or more vectors to give a (multi)dimensional table of the statistic. Overall summary statistics are produced if no classifying vectors are specified. The output vector of the statistic tabulated can be used in further analysis. The syntax is:

$tabulate [**option list**] [**phrase**]s

The input and output parameters of TABULATE are specified in **phrases** by the use of the phrasewords: THE, FOR, WITH, INTO, BY and USING with the following meanings:

THE specifies input vector and statistic
FOR input classification
WITH input weight vector
INTO output table vector
BY output table classification
USING output weight vector

There are suitable defaults when phrases are omitted. The number of possible combinations of input and output parameters is very large so that many types of tabular structure can be produced. If no output phrases are given, the tabular structure is printed. The examples below illustrate the power of this directive.

Examples:

$tabulate the INCOME total

$tabulate the INCOME percentile 90

$tabulate the AGE mean for SEX using BASE into AMEAN

produces mean values of AGE in a vector AMEAN for each level of SEX and the number of observations contributing to each mean in the output vector BASE

$tabulate the AMEAN mean with BASE

lists the overall mean AGE, calculated as a weighted mean of means using weights held in BASE

$tabulate for SEX;CLASS using TABLE by CSEX;CCLASS

forms the two-way SEX by CLASS table of frequencies into an output vector TABLE with the classifying factors output into factors CSEX and CCLASS.

3.12 TPRINT. Prints the values of one or more vectors as a table dimensioned by a set of classifying factors. The layout and labelling of the table can be controlled. The syntax is:

$tprint [**option list**] **list1** [**list2**]

where **list1** specifies the vector(s) to be printed and **list2** the desired shape of the table. The options available are STYLE which controls the edging and annotation of the table and COLS which specifies the maximum width of the table.

Example:

$tprint (style=1) AGE INCOME SEX;CLASS

prints a table, classified by the factors SEX and CLASS, of AGE and INCOME values in each cell, with borders around the table and the rows and columns indexed by the values of the classifying factors.

3.13 TRANSCRIPT. GLIM input and output is now also automatically sent to an external file attached to a GLIM interactive session. This file will normally be a temporary file which can be printed, copied or deleted according to user requirements at the end of the session. The directive can be used to control the type of input and output sent to this file; the user can select/suppress directives input, output from directives, macro verification, warning, fault and help messages.

3.14 VERIFY. A switch to produce/suppress macro verification (each line of an executing macro is output).

4. Changes for Interactive Use

4.1 Prompts for further input now include the directive name if more items may be expected for the directive.

4.2 A soft break-in facility is available on some systems. This allows the user to abandon the directive being executed.

4.3 Some systems prompt for filenames whenever channel numbers are initially referenced in INPUT and OUTPUT directives.

5. Other Enhancements

5.1 The list of systems scalars has been extended. These cover switch settings, input and output channel parameters, seeds for random number generation, additional quantities for model fitting such as error and link settings and the values of the items in the CYCLE/RECYCLE directives and settings for argument usage in macros. The range of system scalars allow macros of complete generality to be written.

5.2 A library of pre-defined macros is now available for use in GLIM 3.77. The first release of the Library includes macros to perform the following functions:

calculation of a set of summary statistics for a vector
testing the distribution of a variate against the Normal and Uniform distributions
stem and leaf plots
Box-Cox transformations
quantile-quantile probability plots
t-values for parameter estimates
Chi-squared probability
Calculation of leverage values after fitting a model with a Normal error distribution
predicted error sum of squares and R-squared for a Normal model
and fitting the Exponential and Weibull distributions to right-censored survival data with plotting of residuals.

5.3 Run-time error detection has been improved. Invalid channel numbers and FORTRAN formats, premature end-of-file etc are now detected and notified by a fault message. On some computers the operating system will prompt for a valid alternative, if appropriate.

5.4 There are now some additional warning messages. Fault messages are now output in textual form rather than as fault numbers.

Acknowledgements

We would like to thank the other members of the GLIM working party of the Royal Statistical Society who have developed and documented GLIM 3.77. Major contributors are Mike Clarke, Brian Francis, John Nelder, Alan Reese, Tony Swan, Roger White and particularly, Robert Baker and Mick Green. Thanks are also due to the many people who have implemented and tested GLIM 3.77 on their machines.

References

Baker, R J and Nelder, J A (1978) The GLIM System Release 3, Numerical Algorithms Group, Oxford.

NEARLY LINEAR MODELS USING GENERAL LINK FUNCTIONS

J.H. Roger
Department of Applied Statistics
University of Reading
Whiteknights, Reading RG6 2AN

SUMMARY

The major advantage of GLIM when fitting Generalised Linear Models is the ability to fit a succession of models by simply redefining the model statement. As such it is a major tool for data exploration. Several authors have used the facilities in GLIM to programme other problems which do not fall into the GLM class. However GLIM is not a good programming language. The successful attempts can be categorised by the fact that they allow the user this same flexibility to alter the model at the subsequent stages of the analysis.

There exists a large number of data analysis problems which satisfy the usual criteria for a GLM apart from a few extra nuisance parameters involved in the link function. The link function is *nearly linear*. Here we describe an approach to this class of problem which can be implemented simply on GLIM and which allows the linear component of the model to be altered using the standard GLIM model definition features.

Two examples are given. One fits a probit model with extra parameters for natural responsiveness. The other fits an inverse polynomial response function with two variables.

BACKGROUND

In the standard Generalised Linear Model a set of n univariate observations y_i are assumed to be a realisation of n independent random variables whose distributions are specific members of the exponential family with associated mean values μ_i. The individual values μ_i are determined by $\mu_i = h(\eta_i)$ where h is any fixed monotonic function of η_i which can vary in form from observation to observation but generally does not. The values of η_i are modelled by a linear combination $X\beta = \Sigma_j X_{ij} \beta_j$ of unknown parameters β_j with fixed design matrix X of elements X_{ij}. The maximum likelihood estimator for β is found in GLIM by iteratively reweighted least squares using the formula $\beta^* = (X' W X)^{-1} X' W z$. The matrix of weights is a diagonal matrix $W = H V^{-1} H$, where V is a diagonal matrix with elements V_{ii} being the variance of the i'th observation. H is a diagonal matrix of elements $H_{ii} = d\mu_i/d\eta_i$, defined by the function h. The working vector z is defined by $z = \eta + H^{-1} (y-\mu)$. When using the *OWN* model facility in GLIM the particular GLM is defined by supplying macros which specify the vectors %FV as $\mu_i = h(\eta_i)$, %DR as $1/H_{ii}$, %VA as V_{ii} from the values of the linear predictor %LP which is η_i. The individual deviance terms need to be defined in the vector %DI, but this is only used when assessing convergence.

The method called composite link functions by Thompson & Baker (1981) allows the mean values μ_i to be modelled by *any* function of the parameters β_i, not merely a general function of a linear combination of the parameters β, say $\mu_i = h_i(\beta)$. It can be shown that the equivalent iteratively reweighted least squares equations for β have the same form $\beta^* = (X' W X)^{-1} X' W z$, where H becomes a unit matrix $H = I_n$ and X becomes the matrix of differentials $X = d\mu/d\beta$ with elements $X_{ij} = d\mu_i/d\beta_j$. This is effectively a linearisation of the response function $\mu_i = h_i(\beta)$. It can theoretically be fitted on GLIM by redefining the design matrix X at each iteration. However this is usually too messy to be a practical computing solution. Thompson and Baker (1981) introduce the concept of a fixed set of linear combinations of the parameters $\eta = X^O \beta$ defined by a fixed design matrix X^O. They then introduce the idea of a matrix C such that $\mu = C h(\eta)$ where h is a monotonic function. The working design matrix becomes $X = C H X^O$ where H is the diagonal matrix of elements $H_{ii} = dh(\eta_i)/d(\eta_i)$. This merely complicates the issue while limiting possible applications. The introduction of the matrix C is difficult to handle using GLIM's limited mathematical facilities. Roger (1982) looks at a class of models involving both sums and products by adding a simple branching structure to the data. The important feature is that the design is defined by the data and not by the macros. Submodels are defined by excluding these terms from the model statement in the usual way.

In fact the development of GLIM algorithms to handle any general problem is not difficult. One must merely redefine the values X_{ij} in the design matrix X at each iteration to take the values $d\mu_i/d\beta_j$ in the same way that V_{ii} is reset at each iteration. The major problem is making the individual GLIM programs cover as many models as possible. For a single parameter β_j the associated variate vector is set to have the value $d\mu/d\beta_j$. When each unit possesses only one of a set of parameters in the model we may use a factor. For the factor A, say, we introduce a variate AX whose i'th value is $d\mu_i/d\beta_j$ where β_j is the parameter associated with the level of A for the i'th observation. Then we include the term A.AX in the fitted model, so that the variate AX weights the factor A. The application of this idea has allowed the fitting of several non-linear models such as genotype-environment interaction models (Roger, to be published). The introduction of factors solves the problem of proliferation of vectors in the model - one variate for each β_j. In GLIM there is a limit on the number of vectors in the model, determined by the number of bits per integer. However this general approach is limited by the fact that the definition of the model is in terms of the macros and not in terms of the data, and hence the model statement, as it is in the approach suggested by Roger (1982).

THEORY

There are several classes of problem where the component of the model, which the user wishes to vary in the link function is a linear function. However extra nuisance parameters, which appear in a standard form, render the total model non-linear. The approach used by several authors is to fit the linear component of the model using GLIM while iterating over the nuisance parameters in some way to obtain a maximum likelihood solution. The variances for the parameter estimates given by GLIM when this is done will of course be underestimates, as they assume the nuisance parameters are fixed. The alternative approach used here will give asymptotically correct standard errors.

As an example consider the standard Probit Analysis problem where $E[Y] = m\Phi(X\beta)$. This can be fitted simply on GLIM using a Probit link function and Binomial Error. Using simple FIT statements a whole series of possible models can be tried such as parallel probits. Finney (1971) suggests an extension to the probit model to allow for natural responsiveness, commonly known as *Abbots formula*, where $E[Y] = m\{C+(1-C)\Phi(X\beta)\}$. This can be further generalised to $E[Y] = m\{C+(1-C-D)\Phi(X\beta)\}$ where C and D are unknown parameters. The value of C represents the response to a minimum level and (1-D) to a maximum level of the linear predictor. The total number tested is m. This example will be used to demonstrate the method. A further application of this approach involving inverse polynomials is also given.

Consider any problem where $\mu_i = h(\Theta,\eta_i)$ where Θ is a vector of q parameters Θ_s and $\eta = X^0\beta$ is the linear component of the model. It is assumed that all the necessary models involve changes to the fixed design matrix X^0. The working design matrix X will involve p parameters β_1 to β_p and the q parameters Θ_1 to Θ_q. For the parameter β_j the i'th value in the working design matrix will be $\{dh(\Theta,\eta_i)/d\eta_i\} X^0_{ij}$. Thus every element in the i'th row of the design matrix X^0 needs to multiplied by the same constant $dh(\Theta,\eta_i)/d\eta_i$ at each iteration. Now this can be done very simply by putting the values $dh(\Theta,\eta_i)/d\eta_i$ into the diagonal elements of the matrix H instead of unit values, where $\beta^* = (X'H\,V^{-1}\,H\,X)^{-1}\,X'H\,V^{-1}\,H\,(X\beta + H^{-1}(y-\mu))$. The elements in the working design matrix X associated with the parameters θ, say X^1, then have terms $\{dh(\theta,\eta_i)/d\theta_s\}/\{dh(\theta,\eta_i)/d\eta_i\}$. Practical computing problems may occur when the denominator in this expression becomes close to zero. Evasive action may be necessary. The conclusion from this is that any model of the form $\mu = h(\theta,\eta)$ where $\eta = X^0\beta$ can be fitted using GLIM in the following way using the *OWN* facility. The basic steps are layed out with cross-reference to the GLIM code that follows.

1. Set %FV to have the appropriate values $h(\theta,\eta)$ based on the linear predictor %LP which has the value $X(\theta',\beta')' = X^1\theta+X^0\beta = X^1\theta+\eta$. So we can calculate η = %LP - $X^1\theta$ where X^1 is the previous working design matrix for the θ parameters. The values for θ must be extracted from the parameter vector %PE. Note that this pre-empts the problem of calculating $X^0\beta$ for any linear design X^0 which the user may specify.

Example: (Section G and following lines)

In our extended Probit example we define two nuisance parameters C and D with GLIM scalar values %C and %D which we extract from %PE. These are associated with variates CC and DD in the model term. In the actual GLIM code PARA(POSI(2)) replaces %C and PARA(POSI(3)) replaces %D. The £FIT statement defines a model including both CC and DD along with the user model.

At each iteration we calculate η as CLP = %LP - CC*%C - DD*%D.
Then %FV = (%C+(1-%C-%D)*%NP(CLP))*M defining the vector
$\mu_i = m_i\{C+(1-C-D)\Phi(X_i^0\beta)\}$

2. Set the vector %DR to have the values $\{H_{ii}\}^{-1} = \{dh(\theta,\eta_i)/d\eta_i\}^{-1}$

Example: (Sections J3 to J6)

As $h(C,D,\eta_i) = m_i\{C+(1-C-D)\ \Phi(\eta_i)\}$ we have
$dh(C,D,\eta_i)/d\eta_i = m_i(1-C-D)\phi(X_i\beta)$.
So %DR = %SQRT(2*%PI)/(M*(1-%C-%D)*%EXP(-CLP*CLP/2))

3. Set the design vectors in X^1 to have values $\{dh(\theta,\eta_i)/d\theta\}/H_{ii}$

Example: (Section J7 to J8)

As $dh(C,D,\eta_i)/dC = m_i\{1-\Phi(\eta_i)\}$ and $dh(C,D,\eta_i)/dD = -m_i\Phi(\eta_i)$ we have the revised design vectors CC and DD for the next iteration,
CC = M*(1-%NP(CLP))*%DR and
DD = -M*%NP(CLP)*%DR

4. Redefine the working vector %LP for the next iteration to allow for the changes in the design matrix. It takes the value $X^1\theta + \eta$ where X^1 is the new value and $\eta = X^0\beta$ is evaluated in section 1 and remains the same since X^0 is not altered.

Example: (Section J9 and following lines)

Using the values of CC and DD from section 3 we reset %LP to the value %LP = CC*%C + DD*%D + CLP where CLP is the value for η which is calculated in section 1.

5. The vector %VA is set to have the values of the variance function in the standard fashion for an exponential family member. (Section K).

6. The vector %DI is defined to have the elemental values for the Deviance statistic. (Section L).

NUMERICAL PROPERTIES

In order to make the iterative solution converge to a solution it is necessary to limit the step size at several stages. This is done in a very crude fashion so that any step which goes down hill (the likelihood decreases), is halved in length until an upward step is found. The halving of the step length is easily obtained by averaging the values of both %PE and %LP with their previous value. More sophisticated stepping procedures seem unnecessary. The initial values for %LP are obtained by fitting the linear component $X^0\beta$ alone for fixed initial values for the extra parameters using standard GLM techniques. (Sections A, E and F).

One possible problem occurs when the values of $(X^0\beta)_i$ are very small compared to $(X^1\theta)_i$ in absolute value, since $(X^0\beta)$ is calculated indirectly by differencing and as such could be subject to rounding error. When this occurs the value of %DR is large. This indicates that the exact value of the linear component for this unit is not important. So the value from the previous iteration is used instead.

USAGE OF MACROS

The macros once defined require initial values for the extra parameters (C and D in the Probit example). If the model does not involve the grand mean %GM then the first element SET(1) of the vector SET should be given the value 0. When one or other of the extra parameters are not being estimated (should be left with there initial values), the values of SET(2), SET(3) etc should be assigned to zero. The vector SET is the unit vector by default, so that the model includes all the extra parameters as well as the grand mean.

The required model is placed in the macro MOD and the macro DROP is normally left blank. When some of extra parameters are not being estimated the macro DROP contains a term such as -CC or -DD to exclude this variable from the model. (See first example).

APPLICATIONS

Two specific examples of the GLIM code are given. The first demonstrates the extended Probit model where three models are fitted. The first is a straight single probit with extra parameters C and D. The second is a parallel Probit model. The third is the same parallel Probit model but with the C parameter fixed to have the value zero.

The second example fits a two variable inverse polynomial response curve. The model is $E[y] = (a+x_1)*(b+x_2)/(c+dx_1+ex_2$ + other higher order terms).

The denominator forms the linear part of this model while a and b are the two extra parameters. The data are from a 4^2 factorial response trial of grass to four levels of Nitrogen and four levels of Potash, with two replicates. A whole series of inverse polynomial models can be fitted by simply altering the model held in the macro MOD.

CONCLUSION

The sets of macros described here solve two specific problems involving the nearly linear link functions. They are not designed to be optimal for either application but rather form an opaque demonstration of the general procedure. It is a trivial task to alter these macros to cover other nearly linear link functions involved in problems with exponential family distributions. The lines of code which need amending are clearly marked. Possible applications include logistic models with an incomplete range of possible values $E[Y] = C+(1-C-D)\ \mathrm{logit}(X\beta)$, further inverse polynomial models and responsiveness models such as $E[Y] = a_i+b_i(X\beta)$. The main feature is that changes in model structure merely involve redefining the model term as specified in the macro MOD. Also the extra parameters can be held constant quite simply at specific fixed values to provide further possible submodels.

REFERENCES

Finney (1971) Probit Analysis. Cambridge

Roger J.H. (1982) Composite Link functions with Linear Log Link and Poisson error. GLIM Newsletter, December 1982.

Thompson, R. & Baker, R.J. (1981). Composite Link functions in generalised Linear Models. Appl Statist. 30, 125-131.

```
!=========    First example                        ====================
!   Probit analysis with extra C and D parameters.
!   Those sections marked with stars are specific to this Example.
!
$MAC CDRUN !
$OWN S1 S2 M3 M4 $!          Set up GLM with fixed extra parameters.
$SCALE 1 $!                                                ************
!
!*** Define initial %LP estimates for FIT with FIXED values for    *A1
!***    the extra parameter values.                                *A2
$CAL %LP=%C+(%YV/M-%C)/(1-%C-%D) $!                       *********A3
  :  %LP=%IF(%GT(%LP,0.99),0.99,%LP) $!                   *********A4
  :  %LP=%IF(%GT(%LP,0.01),%LP,0.01) $!                   *********A5
  :  %LP=%ND(%LP) $!                                      *********A6
!
$OUT 0 $!
$FIT £MOD $!     Fit the model with fixed values for extra parameters.
$OUT %POC $!         to get starting value for %LP.
!
!    Delete working vectors in case previous try failed.
$DEL PARA PP KPE TEMQ CLP PALP KCLP POSI $!
!    The following section deals with the control of extra parameters.
!    Their presence in the model is determined by SET(2) SET(3) etc.
!    SET(1) indicates presence or absence of Grand Mean in the model.
$VAR %PL PP $!
$CAL %Z9=%CU(SET)-SET(1) : PP=%GL(%PL,1)+%Z9 $!
  :  PP(1)=%IF(SET(1),1,1+%Z9) $!
$EXT %PE $!
$CAL %Z9=%PL+%Z9 $VAR %Z9 PARA KPE $!
$CAL PARA(PP)=%PE : POSI=%CU(SET) $!
!
!*** Put user's initial estimates of extra parameters into PARA    *B1
!*** Also initialise the vectors which represent extra parameters. *B2
$CAL PARA(POSI(2))=%C : PARA(POSI(3))=%D $!               *********B3
$CAL DD=0 : CC=0 $!                                       *********B4
!
$CAL %Z5=1.0E50 $!          Set large initial value for deviance.
  :  %Z3=0 : %Z1=1 $!       Set initial values for flags.
  :  KCLP=%LP $!            Set KCLP with initial value of %LP.
$OWN M1 M2 M3 M4 $!
$SCALE 1 $!                                                ************
!
!*** Specify extra parameter vectors at start of FIT statement     *C1
$FIT CC+DD+£MOD £DROP   $DISP E $!                        *********C2
!
!    Tidy up the working vectors.
$DEL PARA PP KPE TEMQ CLP PALP KCLP POSI $!
!
!*** Delete the working vectors in this particular example.        *D1
$DEL TEMP CNP $!                                          *********D2
!
$ENDM!

$MAC S1 $!
!*** Macro for GLM with fixed values of extra parameters.           *E1
$CAL %FV=M*(%C+(1-%C-%D)*%NP(%LP)) $!                     *********E2
!
$ENDM !

$MAC S2 !
!*** Macro for GLM with fixed values of extra parameters.           *F1
$CAL %DR=%SQRT(2*%PI)/(M*(1-%C-%D)*%EXP(-%LP*%LP/2)) $!   *********F2
!
$ENDM !
```

```
$MAC M1 !
$SWI %Z3 GECD $! Extract parameter estimates when not in first loop.
$CAL %Z2=0 $!
$WHI %Z1 WORK $!              Iterate until we have a legal step.
$CAL %Z1=1 : %Z3=1 $!         Reset flags for next iteration.
  :  KPE=PARA : %Z5=%Z4 $!    Keep Parameter estimates and Deviance for
  :  KCLP=CLP $!                  possible use in STEP in next iteration.
$ENDM!                        Keep CLP in case of numeric problems.

$MAC GECD !
$EXT %PE $CAL PARA=%PE $!
$ENDM!

$MAC WORK !
!
!*** Calculate PALP, the part of %LP due to extra parameters.      *G1
$CAL PALP=PARA(POSI(2))*CC+PARA(POSI(3))*DD $!            *********G2
!
$CAL CLP=%LP-PALP $!          Put linear part into CLP.
!    If ABS((%LP-PALP)/(%LP+PALP)) is small then we have potential for
!     rounding error. Set CLP to its previous value at last iteration.
  :  TEMQ=CLP/(%LP+PALP) $!
  :  TEMQ=%IF(%GT(TEMQ,0),TEMQ,-TEMQ) $!
  :  CLP=%IF(%GT(TEMQ,1.0E-6),CLP,KCLP) $!
!
!*** Calculate %FV from the values in CLP and other parameters.    *H1
$CAL CNP=%NP(CLP) $!                                      *********H2
$CAL %FV=(PARA(POSI(2))+ !                                *********H3
           (1-PARA(POSI(2))-PARA(POSI(3)))*CNP)*M $!      *********H4
!
!*** Do any validation checks here and set flag %Z1 appropriately. *I1
$CAL %Z1=%GT(%CU(%LT(%FV,0)+%GT(%FV,M)),0) $!Check binomial*********I2
!
$USE M4 $!
$CAL %Z4=%CU(%DI) $!
  :  %Z1=%GT(%Z1+%GT(%Z4,%Z5+0.001),0) $!  Check if going up hill.
$SWI %Z1 STEP $!                           If error half step length.
$ENDM!

$MAC STEP !       This macro reduces the step length by half.
$CAL %Z9=%GT(%Z2,20) $SWI %Z9 ERR1 $!
$CAL %LP=(%LP+KLP)/2 : PARA=%PE=(%PE+KPE)/2 : %Z2=%Z2+1 $!
$ENDM!

$MAC ERR1 !
$PRI 'MORE THAN 20 BACKWARD STEPS IN ALGORITHM' $!
  :  'PROBLEM IN CONVERGANCE' $!
$DEL PARA PP KPE TEMQ CLP PALP KCLP POSI $!
$YVAR %YV $!      This is a way to escape safely from OWN in GLIM377!!
$ENDM!              (in GLIM3 use $CALC %F=%CL-1 $EXIT %F$ instead)

$MAC M2 !
!
!*** Calculate %DR and vectors defining extra parameters.          *J1
!*** Set PALP as the part of %LP represented by extra parameters.  *J2
$CAL TEMP=%LOG(%SQRT(2*%PI)/ !                            *********J3
        (M*(1-PARA(POSI(2))-PARA(POSI(3)))))+CLP*CLP/2 $! *********J4
$CAL TEMP=%IF(%GT(TEMP,100),100,TEMP) $!                  *********J5
$CAL %DR=%EXP(TEMP) $!                                    *********J6
```

```
$CAL CC=M*(1-CNP)*%DR $!                                        *********J7
$CAL DD=-M*CNP*%DR $!                                           *********J8
$CAL PALP=CC*PARA(POSI(2))+DD*PARA(POSI(3)) $!                  *********J9
!
$CAL  %LP=CLP+PALP $!
  :   KLP=%LP $!  Save linear predictor in case we need to step back.
$ENDM!

$MAC M3 !
!***           Define Mean-Variance relationship for this ERROR     *K1
!              Binomial Error.                                  *********K2
$CAL %VA=%FV*(M-%FV)/M!                                         *********K3
!
$ENDM!

$MAC M4 !
!***         Define Deviance for this ERROR                             *L1
!            Binomial Error.                                    *********L2
$WARN $!     Turn Warnings Off and On in case %YV is zero.      *********L3
$CAL %DI=2*(%YV*%LOG(%YV/%FV)+(M-%YV)*%LOG((M-%YV)/(M-%FV))) $!*****L4
  :  %Z9=%CU(%DI) : %DI=%IF(%GT(%Z9,0),%DI,0) $!                *********L5
$WARN $!                                                        *********L6
!
$ENDM!

!**  Length of SET is (number of extra parameters +1)                  *M1
$VAR 3 SET $!                                                   *********M2
$ASSIGN SET=1,1,1 $! Defualt values in vector SET.              *********M3

$MAC DROP  $ENDM !            Default blank macro for DROP.

!=========     First example application              ====================
$UNITS 22$
$FACTOR GROUP 2$
$DATA RESP M    GROUP DOSE $READ
16  100  1  -10               15  100  2  -10
14  100  1  -5                18  100  2  -5
16  100  1  3                 35  100  2  3
18  100  1  4                 40  100  2  4
25  100  1  5                 52  100  2  5
33  100  1  6                 58  100  2  6
38  100  1  7                 55  100  2  7
40  100  1  8                 62  100  2  8
49  100  1  9                 66  100  2  9
61  100  1  10                72  100  2  10
70  100  1  20                72  100  2  20

$PLOT RESP DOSE 'AB' GROUP
$YVAR RESP

$CAL %C=%D=0 $              ! Set initial values for C and D parameters.

$MACRO MOD  DOSE           $ENDM $USE CDRUN$     ! One line.
$MACRO MOD  GROUP+DOSE     $ENDM $USE CDRUN$     ! Parallel lines.

$CALC SET(2)=0 : %C=0 $   ! Fix the C parameter at 0 for this FIT.
$MACRO DROP -CC            $ENDM $USE CDRUN$

$STOP
```

```
!=========    Results from first example         ====================

 76.00 *
 72.00 *                                    B               2
 68.00 *                                  B
 64.00 *                                B
 60.00 *                             B      A
 56.00 *                               B
 52.00 *                           B
 48.00 *                                  A
 44.00 *
 40.00 *                          B    AA
 36.00 *                        B
 32.00 *                             A
 28.00 *
 24.00 *                           A
 20.00 *           B              A
 16.00 *  2        A            A
 12.00 *
  8.00 *
  4.00 *
  0.00 *
......*.........*.........*.........*.........*.........*.........*
   -12.0      -6.0       0.0       6.0      12.0      18.0      24.0

-- model changed
deviance =  78.94 at cycle  6
    d.f.=  18

         estimate         s.e.     parameter
    1      -1.718       0.2445     1
    2      0.1566      0.01824     CC
    3      0.2860      0.03109     DD
    4      0.2763      0.03950     DOSE
    scale parameter taken as  1.000

-- model changed
-- model changed
deviance =  6.516 at cycle  6
    d.f. = 17

         estimate         s.e.     parameter
    1      -2.585       0.3944     1
    2      0.1520      0.01735     CC
    3      0.2928      0.02693     DD
    4       1.205       0.2102     GROU(2)
    5      0.3267      0.05248     DOSE
    scale parameter taken as  1.000

-- model changed
-- model changed
deviance = 58.570 at cycle  2
    d.f. = 18

         estimate         s.e.     parameter
    1     -0.6753      0.08577     1
    2     0.07393      0.09750     DD
    3      0.4261      0.06368     GROU(2)
    4     0.06609     0.006660     DOSE
    scale parameter taken as  1.000
```

```
!=========    Second Example                        =====================
!   Bivariate Inverse polynomial.
!   Those sections marked with stars are specific to this Example.
!
$MAC IPRUN !
$OWN S1 S2 M3 M4 $!          Set up GLM with fixed extra parameters.
!
!*** Define initial %LP estimates for FIT with FIXED values for     *A1
!***    the extra parameter values.                                 *A2
$CAL %LP=(%A+X1)*(%B+X2)/%YV $!                           *********A3
!
$OUT 0
$FIT £MOD $!      Fit the model with fixed values for extra parameters.
$OUT %POC $!         to get starting value for %LP.
!
!    Delete working vectors in case previous try failed.
$DEL PARA PP KPE TEMQ CLP PALP KCLP POSI $!
!    The following section deals with the control of extra parameters.
!    Their presence in the model is determined by SET(2) SET(3) etc.
!    SET(1) indicates presence or absence of Grand Mean in the model.
$VAR %PL PP $!
$CAL %Z9=%CU(SET)-SET(1) : PP=%GL(PL,1)+%Z9 $!
  :  PP(1)=%IF(SET(1),1,1+%Z9) $!
$EXT %PE $!
$CAL %Z9=%PL+%Z9 $VAR %Z9 PARA KPE $!
$CAL PARA(PP)=%PE : POSI=%CU(SET) $!
!
!*** Put user's initial estimates of extra parameters into PARA     *B1
!*** Also initialise the vectors which represent extra parameters. *B2
$CAL PARA(POSI(2))=%A : PARA(POSI(3))=%B $!                *********B3
$CAL AA=0 : BB=0 $!                                        *********B4
!
$CAL %Z5=1.0E50 $!          Set large initial value for deviance.
  :  %Z3=0 : %Z1=1 $!       Set initial values for flags.
  :  KCLP=%LP $!            Set KCLP with initial value of %LP.
$OWN M1 M2 M3 M4 $!
!
!*** Specify extra parameter vectors at start of FIT statement      *C1
$FIT AA+BB+£MOD £DROP   $DISP E $!                         *********C2
!
!    Tidy up the working vectors.
$DEL PARA PP KPE TEMQ CLP PALP KCLP POSI $!
!
!*** Delete the working vectors in this particular example.         *D1
!
$ENDM!

$MAC S1 !
!*** Macro for GLM with fixed values of extra parameters.            *E1
$CAL %FV=(%A+X1)*(%B+X2)/%LP $!                            *********E2
!
$ENDM !

$MAC S2 !
!*** Macro for GLM with fixed values of extra parameters.            *F1
$CAL %DR=-%LP/%FV $!                                       *********F2
!
$ENDM !
```

```
$MAC M1 !
$SWI %Z3 GECD $! Extract parameter estimates when not in first loop.
$CAL %Z2=0 $!
$WHI %Z1 WORK $!              Iterate until we have a legal step.
$CAL %Z1=1 : %Z3=1 $!         Reset flags for next iteration.
  :  KPE=PARA : %Z5=%Z4 $!   Keep Parameter estimates and Deviance for
  :  KCLP=CLP $!                 possible use in STEP in next iteration.
$ENDM!                        Keep CLP in case of numeric problems.

$MAC GECD !
$EXT %PE $CAL PARA=%PE $!
$ENDM!

$MAC WORK !
!
!*** Calculate PALP, the part of %LP due to extra parameters.       *G1
$CAL PALP=PARA(POSI(2))*AA+PARA(POSI(3))*BB $!               *********G2
!
$CAL CLP=%LP-%PALP $!          Put linear part into CLP.
!    If ABS((%LP-PALP)/(%LP+PALP)) is small then we have potential for
!     rounding error. Set CLP to its previous value at last iteration.
  :  TEMQ=CLP/(%LP+PALP) $!
  :  TEMQ=%IF(%GT(TEMQ,0),TEMQ,-TEMQ) $!
  :  CLP=%IF(%GT(TEMQ,1.0E-6),CLP,KCLP) $!
!
!*** Calculate %FV from the values in CLP and other parameters.     *H1
$CAL %FV=(PARA(POSI(2))+X1)*(PARA(POSI(3))+X2)/CLP $!        *********H2
!
!*** Do any validation checks here and set flag %Z1 appropriately. *I1
$CAL %Z1=0 $!                                                 *********I2
!
$USE M4 $!
$CAL %Z4=%CU(%DI) $!
  :  %Z1=%GT(%Z1+%GT(%Z4,%Z5+0.001),0) $!  Check if going up hill.
$SWI %Z1 STEP $!                            If error half step length.
$ENDM!

$MAC STEP !       This macro reduces the step length by half.
$CAL %Z9=%GT(%Z2,20) $SWI %Z9 ERR1 $!
$CAL %LP=(%LP+KLP)/2 : PARA=%PE=(%PE+KPE)/2 : %Z2=%Z2+1 $!
$ENDM!

$MAC ERR1 !
$PRI 'MORE THAN  20 BACKWARD STEPS IN ALGORITHM' $!
  :  'PROBLEM IN CONVERGANCE' $!
$DEL PARA PP KPE TEMQ CLP PALP KCLP POSI $!
$YVAR %YV $!      This is a way to escape safely from OWN in GLIM377!!
$ENDM!             (in GLIM3 use $CALC %F=%CL-1 $EXIT %F$ instead)

$MAC M2 !
!
!*** Calculate %DR and vectors defining extra parameters.            *J1
!*** Set PALP as the part of %LP represented by extra parameters.  *J2
$CAL %DR=-CLP/%FV $!                                          *********J3
$CAL AA=-(PARA(POSI(3))+X2)/%FV $!                            *********J4
$CAL BB=-(PARA(POSI(2))+X1)/%FV $!                            *********J5
$CAL PALP=PARA(POSI(2))*AA+PARA(POSI(3))*BB $!               *********J7
!
$CAL %LP=CLP+PALP $!
  :  KLP=%LP $!  Save linear predictor in case we need to step back.
$ENDM!
```

```
$MAC M3 !
!***          Define Mean-Variance relationship for this ERROR    *K1
!             Normal   Error.                                *********K2
$CAL %VA=1 $!                                                *********K3
!
$ENDM!

$MAC M4 !
!***          Define Deviance for this ERROR                        *L1
!             Normal   Error.                                *********L2
$CAL %DI=(%YV-%FV)*(%YV-%FV) $!                              *********L3
!
$ENDM!

!**  Length of SET is (number of extra parameters +1)              *M1
$VAR 3 SET $!                                                *********M2
$ASSIGN SET=1,1,1 $! Defualt values in vector SET.           *********M3

$MAC DROP $ENDM !          Default blank macro for DROP.
!=========     Second example application          ====================
$UNITS 32$
$DATA RESP $READ
 4.63  4.40  6.77  6.74  5.70  6.34  5.51  5.30
 5.91  5.42 11.53 11.34 11.01 11.07  9.62  8.95
 5.47  4.85 13.07 12.92 12.84 12.86 10.90 11.37
 4.28  4.50 13.12 11.90 13.28 14.22 11.84 12.40
$CAL X1=(%GL(4,2)-1)*200 : X2=(%GL(4,8)-1)*200
$CAL X11=X1*X2 : X20=X1*X1 : X21=X20*X2
$YVAR RESP
     ! Set initial values for intercept parameters.
$CAL %A=100 : %B=100
$MACRO MOD  X1+X2+X20           $ENDM $USE IPRUN$
$MACRO MOD  X1+X2+X20+X11+X21 $ENDM $USE IPRUN$
$STOP
!=========     Results from second example         ====================
 -- model changed
 deviance = 20.688 at cycle  4
     d.f. = 26

          estimate          s.e.      parameter
     1        4810.         1536.     1
     2        89.47         26.48     AA
     3        252.2         55.34     BB
     4        6.857         10.08     X1
     5        16.72         3.653     X2
     6      0.07040       0.01152     X20
     scale parameter taken as  0.7957

-- model changed
-- model changed
deviance =  5.9403 at cycle  3
    d.f. = 24

          estimate          s.e.      parameter
    1        1352.         479.6      1
    2        81.57         24.41      AA
    3        77.51         14.80      BB
    4        6.123         1.981      X1
    5        15.98         4.650      X2
    6      0.01460      0.004225      X20
    7    -0.003224       0.01989      X11
    8  0.00009220   0.00002756       X21
    scale parameter taken as  0.2475
```

IMPLEMENTATION OF AN ALGORITHM FOR FITTING A CLASS OF GENERALIZED LOGISTIC MODELS

Thérèse Stukel
Dartmouth Medical School, Hanover, New Hampshire 03756

SUMMARY

A general class of models is proposed for describing the dependence of binary data on explanatory variables, extending the scope of the standard logistic model. A simple and intuitive algorithm for fitting parameters is presented, based on the properties of generalized linear models; implementation is facilitated through the flexibility of GLIM. Asymptotic variance expressions are derived for the parameters and the fitted values. User defined GLIM macros for fitting the model are included and an example is presented.

Keywords: binary response, generalized linear model, GLIM, goodness of fit, logistic model, logit, probit, shape parameter.

1. INTRODUCTION

The linear logistic model is widely used for modelling the dependence of binary data on explanatory variables. Formally, this model is a generalized linear model (g.l.m.) with a Bernoulli response whose mean μ is expressed as a function of the explanatory variables, $\mathbf{x}' = (x_1, \ldots, x_k)$, in the form

$$\mu(\eta) = \frac{\exp(\eta)}{1+\exp(\eta)} \quad \text{or} \quad \log \frac{\mu(\eta)}{1-\mu(\eta)} = \eta = \mathbf{x}'\boldsymbol{\beta}$$

where $\boldsymbol{\beta} \in \mathbf{R}^k$ are the parameters of interest. A thorough account of the techniques and the motivation for the standard logistic model is presented in Cox (1970) and Nelder and Wedderburn (1972).

This model is not, however, without its limitations. The functional form of $\mu(\eta)$ gives rise to some of the restrictions of the logistic model. For any value of $\boldsymbol{\beta}$, the response curve has the fixed form described above which is skew-symmetric about $\eta = 0$. Data which are not skew-symmetric, or are skew-symmetric but have a steeper or gentler incline in the central probability region may not be well fitted by this model. Furthermore, the maximum likelihood estimation procedure weights each observation according to its estimated (Bernoulli) variance $\mu(\eta)\cdot(1-\mu(\eta))$ so that points in the central probability region near $\mu = 1/2$ have the strongest influence on the fit.

The proposed extension uses two parameters which specifically influence the behaviour of the fitted curve on the tails while allowing for adjustments to shape along the rest of the curve.

2. FORMULATION OF THE PROPOSED MODEL

The general form of the proposed model is

$$\mu(\eta) = \frac{\exp(h(\eta))}{1+\exp(h(\eta))}, \quad \eta = \mathbf{x}'\boldsymbol{\beta} \in \mathbf{R}$$

where $h(\eta)$ is a strictly increasing non-linear function of η.

The system of h-functions used is the following (see Figures 1a and 1b):
For $\eta \geqslant 0$ ($\mu \geqslant \frac{1}{2}$),

$$h(\eta) = \begin{cases} \alpha_1^{-1}(e^{\alpha_1|\eta|}-1), & \alpha_1 > 0 \\ \eta, & \alpha_1 = 0 \\ -\alpha_1^{-1}\log(1-\alpha_1|\eta|), & \alpha_1 < 0 \end{cases}$$

and for $\eta \leqslant 0$ ($\mu \leqslant \frac{1}{2}$),

$$h(\eta) = \begin{cases} -\alpha_2^{-1}(e^{\alpha_2|\eta|}-1), & \alpha_2 > 0 \\ \eta, & \alpha_2 = 0 \\ \alpha_2^{-1}\log(1-\alpha_2|\eta|), & \alpha_2 < 0 . \end{cases}$$

These particular functions provide an extension with the required properties. The case $h(\eta)=\eta$ ($\alpha_1=\alpha_2=0$) is the logistic model. Each parameter governs behaviour on one tail of the curve; for example, $\alpha_1 < 0$ gives a curve with a gentler rise than the logistic on the upper tail while positive α_1 gives a steeper curve. When $\alpha_1=\alpha_2$, skew-symmetric curves with varying slopes in the central region are generated. Different α_1 and α_2 produce a variety of symmetric and asymmetric models encompassing a wider range of situations than the standard model.

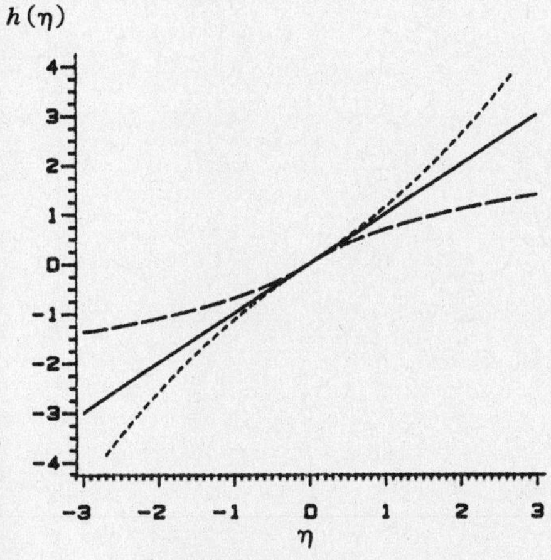

Figure 1a Plot of $h(\eta)$ against η

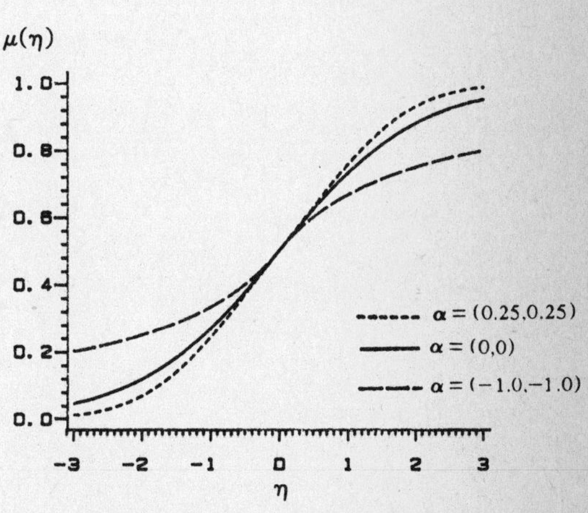

Figure 1b Plot of $\mu(\eta)$ against η

For fixed values of the shape parameter α, the model is a g.l.m. Thus all the power of generalized linear model theory as well as that of the interactive modelling package GLIM can be brought to bear on this extension. In particular, maximum likelihood estimation of the linear parameters using an iterative weighted least squares procedure and calculation of the fitted response values can be accomplished easily with user-defined macros in GLIM. Output from GLIM facilitates estimation of the non-linear shape parameters through calculation of the *deviance* (or twice the drop in log-likelihood from its unrestricted maximum value) for each set of parameter values. Asymptotic variance estimates of the parameters and the fitted values are easily computed using components produced in GLIM.

3. THE FITTING PROCEDURE

The parameters in the model partition neatly into two groups, the linear $\boldsymbol{\beta} \in \mathbf{R}^k$ and the non-linear $\boldsymbol{\alpha} \in \mathbf{R}^2$. This suggests a cyclic procedure for fitting the maximum likelihood estimates. Since for fixed $\boldsymbol{\alpha}$ the model is a g.l.m., GLIM is used to calculate the maximum likelihood estimate of $\boldsymbol{\beta}$ given $\boldsymbol{\alpha}$, $\hat{\boldsymbol{\beta}}_\alpha$, for a set of $\boldsymbol{\alpha}$ values provided by the user. The starting value for each step is the estimated value of $\boldsymbol{\beta}$ from the previous step, beginning with $\hat{\boldsymbol{\beta}}_0$, the maximum likelihood estimate from the standard logistic model. The pair $(\hat{\boldsymbol{\beta}},\hat{\alpha})$ which minimizes the deviance is the maximum likelihood estimate.

Working interactively, it is possible to iterate quickly over values of (α_1, α_2) using two GLIM statements (marked with '!' in the listing at the end). Output from GLIM is used to guide the user in determining the optimum parameter values. Information about the α-values is present in superimposed plots of observed and fitted probabilities at every step of the procedure: for example, if the observed points are rising to the limiting value of 1 faster than the fitted curve on the higher end then α_1 should be increased in the next cycle. Individual deviance values for each observation are also useful in detecting lack of fit and suggesting a direction for further improvement. In practice, this diagnostic information has been useful in guiding the proposed procedure so that it converges with little trouble. Divergence usually indicates that some α is too large so that a fit of 0 or 1 occurs for an observation where the probability is not a hard 0 or 1, causing the deviance to become infinite. A good starting value for the fitting procedure and care in use are needed to ensure convergence.

By taking advantage of the structure and properties of a g.l.m., the optimization in $k+2$ dimensions is reduced to a two-dimensional search for the α's and an efficient k-dimensional maximization over $\boldsymbol{\beta}$. The rate of convergence for the linear parameters is second order since GLIM is implementing a Fisher scoring procedure.

A number of "automatic" optimization procedures could be implemented instead of this "manual" method for finding the maximum over $\boldsymbol{\alpha}$. For instance, it is possible to cycle between the two groups of parameters, maximizing the conditional log-likelihood function over one set of parameters while keeping the other set fixed. At each step, this is accomplished by setting the conditional score function to 0 and solving. For fixed $\boldsymbol{\alpha}$, this system is identical to the maximum likelihood equations used by GLIM for fitting $\boldsymbol{\beta}$ given $\boldsymbol{\alpha}$; for fixed $\boldsymbol{\beta}$, it consists of two non-linear equations in $\boldsymbol{\alpha}$ which can be solved using the Newton-Raphson method. Automatic optimization procedures which maximize conditional log-likelihoods or expected log-likelihoods would work well and require less user input than the proposed method but would require a more complex algorithm and the problem of divergence could still occur.

4. VARIANCE EXPRESSIONS

An estimate of the asymptotic variance matrix of the parameters $(\hat{\boldsymbol{\beta}},\hat{\alpha})'$, $\mathbf{V}_{\hat{\beta},\hat{\alpha}}$, is the inverse of the expected Fisher information matrix. It can be shown that (Stukel, 1983),

$$\mathbf{V}_{\hat\beta,\hat\alpha} = \begin{bmatrix} \mathbf{X'WX} & \mathbf{X'C} \\ \mathbf{C'X} & \mathbf{U} \end{bmatrix}^{-1} = \begin{bmatrix} \mathbf{B}^{-1}+\mathbf{MA}^{-1}\mathbf{M'} & -\mathbf{MA}^{-1} \\ -\mathbf{A}^{-1}\mathbf{M'} & \mathbf{A}^{-1} \end{bmatrix}$$

where $\mathbf{B}^{-1} = (\mathbf{X'WX})^{-1}$ is the variance-covariance matrix of $\hat{\boldsymbol\beta}$ given $\hat{\boldsymbol\alpha}$ as output by GLIM at convergence, $\mathbf{X}$ is the $(n \times k)$ matrix of independent observations, $\mathbf{W}$ is the weight matrix at convergence of the weighted least squares procedure in GLIM, $\mathbf{M} = \mathbf{B}^{-1}\mathbf{X'C}$ is a $(k \times 2)$ matrix, $\mathbf{A}^{-1} = (\mathbf{U} - \mathbf{C'XB}^{-1}\mathbf{X'C})^{-1}$ is the variance matrix of $\hat\alpha$, a (2×2) matrix,

$$\mathbf{C'} = \begin{bmatrix} \mu_i(1-\mu_i)\cdot\frac{\partial}{\partial\eta_i}h_\alpha(\eta_i)\cdot\frac{\partial}{\partial\alpha_1}h_\alpha(\eta_i),\ i=\overline{1,n} \\ \mu_i(1-\mu_i)\cdot\frac{\partial}{\partial\eta_i}h_\alpha(\eta_i)\cdot\frac{\partial}{\partial\alpha_2}h_\alpha(\eta_i),\ i=\overline{1,n} \end{bmatrix} \text{ is a } (2 \times n) \text{ matrix,}$$

and $\mathbf{U} = diag\,\{u_j = \sum_{i \in G_j} \mu_i(1-\mu_i)\cdot\left[\frac{\partial}{\partial\alpha_j}h_\alpha(\eta_i)\right]^2,\ j=1,2\}$.

G_1 is the group of points with a positive value of η and G_2 those with a negative value of η. The macro VCOV in the GLIM listing extracts $\mathbf{B}^{-1}$ and calculates and stores the elements $\mathbf{X}$, $\mathbf{C}$ and $\mathbf{U}$ which are passed to a software package capable of performing the necessary matrix manipulations. Note that $\mathbf{V}_{\hat\beta,\hat\alpha}$ can be calculated by inverting at most a two-dimensional matrix, $\mathbf{A}$.

For most applications, however, it is the estimate of the variance of the fitted curve $\mu(\eta)$ that is of interest. The estimate of $\mu(\eta)$ is a function of $\hat{\boldsymbol\beta}$ and $\hat{\boldsymbol\alpha}$ so that its precision depends on the precision of these parameters. A first-order Taylor expansion of the log odds about its true value is taken since it is more "normal" than μ itself. This gives

$$\log\frac{\hat\mu}{1-\hat\mu} = \hat h \approx h + \sum_{j=1}^{k}(\hat\beta_j - \beta_j)\cdot\frac{\partial}{\partial\beta_j}h + \sum_{j=1}^{2}(\hat\alpha_j - \alpha_j)\frac{\partial}{\partial\alpha_j}h$$

where $\hat\mu = \mu_{\hat\alpha}(\hat\eta)$ and $\hat h = h_{\hat\alpha}(\hat\eta)$. Thus $\mathrm{var}(\log\hat\mu/(1-\hat\mu)) = \mathrm{E}(\hat h - h)^2 \approx \hat{\boldsymbol\delta}'\cdot\mathbf{V}_{\hat\beta,\hat\alpha}\cdot\hat{\boldsymbol\delta}$ where $\hat{\boldsymbol\delta}$ is vector of partial derivatives of h with respect to the parameters, evaluated at $(\hat{\boldsymbol\beta},\hat{\boldsymbol\alpha})$. Finally, an approximate large-sample 95% confidence interval for $\log\mu/(1-\mu)$, calculated as two standard deviations about the maximum likelihood estimate of this quantity, can easily be transformed to give a 95% confidence interval for $\mu(\eta)$.

5. EXAMPLE

Bliss (1935) presents a study of adult beetle mortality after five hours' exposure to gaseous carbon disulphide. Table 1 reports the data consisting of the dosages of CS_2 and the total number of beetles exposed and killed. Standard logistic analysis results in a deviance of 11.23 with 6 degrees of freedom. This, as well as the superimposed plots of observed and fitted probabilities, suggests that improvements might be possible.

Using the h-family, the deviance is lowest when $\alpha_1 = 0.16$ and $\alpha_2 = -0.53$; its value is 2.74 with 4 degrees of freedom. The drop in deviance from the standard logistic model is 8.49 units (p=0.015 $c.f.\ \chi_2^2$). Table 1 gives approximate 95% confidence intervals for the expected numbers of beetles killed at each dosage and Figure 2 shows the observed and fitted probability curves for both analyses, all superimposed. The parameter estimates and the asymptotic covariance matrices for the logistic and h-family models are

$$\hat{\boldsymbol{\beta}} = \begin{pmatrix} -60.72 \\ 34.27 \end{pmatrix} \quad \mathbf{V}_{\hat{\beta}} = \begin{pmatrix} 26.83 & -15.08 \\ -15.08 & 8.48 \end{pmatrix} \quad \text{and} \quad \begin{pmatrix} \hat{\boldsymbol{\beta}} \\ \hat{\boldsymbol{\alpha}} \end{pmatrix} = \begin{pmatrix} -69.94 \\ 39.25 \\ 0.16 \\ -0.53 \end{pmatrix} \quad \mathbf{V}_{\hat{\beta},\hat{\alpha}} = \begin{pmatrix} 524 & -293 & 5.6 & 8.7 \\ -293 & 164 & -3.1 & -4.8 \\ 5.6 & -3.1 & 0.07 & 0.09 \\ 8.7 & -4.8 & 0.09 & 0.20 \end{pmatrix}.$$

h-family confidence intervals may be wider than logistic since the shape parameter contributes additional variation. For the most part, the standard errors do not increase greatly. An increase in prediction error is an acceptable price to pay when the fit has substantially improved.

TABLE 1. ADULT BEETLE MORTALITY AFTER EXPOSURE TO CS_2

Dosage CS_2	Number of Beetles Killed	Exposed	Logistic 95% C.I. for $n\hat{\mu}$	h-family 95% C.I. for $n\hat{\mu}$
1.6907	6	59	2.03, 5.79	3.74, 12.44
1.7242	13	60	7.03, 13.49	7.95, 15.04
1.7552	18	62	18.78, 26.39	14.13, 24.24
1.7842	28	56	30.73, 36.92	23.91, 34.62
1.8113	52	63	46.77, 52.89	43.03, 53.87
1.8369	53	59	50.78, 55.09	51.67, 56.72
1.8610	61	62	57.40, 60.34	58.26, 61.69
1.8839	60	60	57.62, 59.34	56.53, 59.99

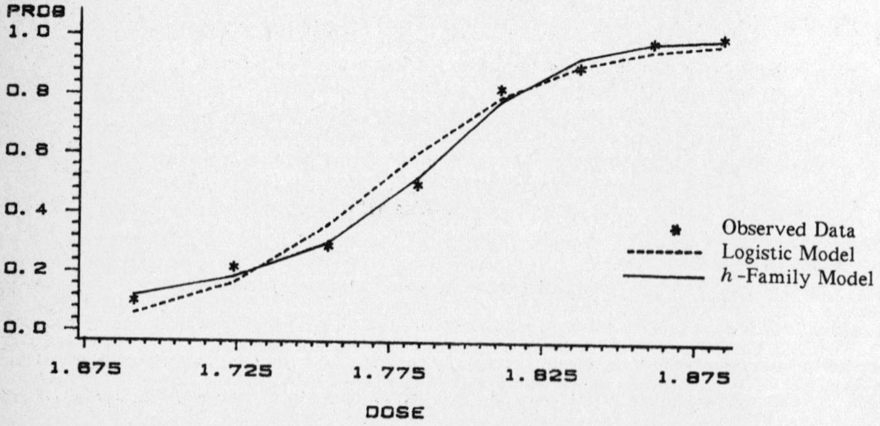

Figure 2 ADULT BEETLE MORTALITY Fitted Probability Curves for the Standard and h-Family Models

LISTING OF GLIM SESSION ILLUSTRATING IMPLEMENTATION OF ALGORITHM

```
$C STATEMENTS ARE INCLUDED TO PREVENT OVERFLOW AND UNDERFLOW
$C                    ON SMALLER INSTALLATIONS
$MACRO M1
$CALC H=%LP
   : %FV=%GE(H,0)*(%GT(%A,0)*(%EXP(%A*H)-1)/%A-
        %LT(%A,0)*%LOG(1-%A*H)/%A+%EQ(%A,0)*H)+
        %LT(H,0)*(%LT(%B,0)*%LOG(1+%B*H)/%B-
        %GT(%B,0)*(%EXP(-%B*H)-1)/%B+%EQ(%B,0)*H)
   : %FV=%IF(%GT(%FV,80),80,%FV)
   : %FV=N*(%EXP(%FV)/(1+%EXP(%FV))) $ENDMAC

$MACRO M2
$CALC %DR=%FV*(N-%FV)/N
   : %DR=%DR*(%GE(H,0)*(%GT(%A,0)*%EXP(%A*H)+
        %LT(%A,0)/(1-%A*H)+%EQ(%A,0))+
        %LT(H,0)*(%LT(%B,0)/(1+%B*H)+
        %GT(%B,0)*%EXP(-%B*H)+%EQ(%B,0)))
   : %DR=%IF(%LT(%DR,%E),%E,%DR)
   : %DR=1/%DR $ENDMAC

$MACRO M3
$CALC %VA=%FV*(N-%FV)/N
   : %VA=%VA*%GT(%VA,%E)+%E*%LE(%VA,%E) $ENDMAC

$MACRO M4
$CALC %DI=2*(Y*%LOG(Y/%FV)+(N-Y)*%LOG((N-Y)/(N-%FV)))
   : H=%EQ(%FV,0)*%GT(Y,0)+%EQ(%FV,N)*%LT(Y,N)
   : %DI=%DI*%EQ(H,0)+(1/%E)*%EQ(H,1) $ENDMAC

$MAC INT
$C INT IS USED FOR INTERACTIVE FITTING OF H-FAMILY MODEL
$FIT DOSE $DISP E $CALC G=%FV/N $PLOT G P DOSE $ENDMAC

$MACRO VCOV
$C VCOV CALCULATES COMPONENTS FOR THE PARAMETER VAR/COV MATRICES
$CALC H=%LP
   : VA=%GE(H,0)*(%GT(%A,0)*(%EXP(%A*H)*(%A*H-1)+1)**2+
             %LT(%A,0)*(%LOG(1-%A*H)+%A*H/(1-%A*H))**2)
   : VB=%LT(H,0)*(%LT(%B,0)*(%LOG(1+%B*H)-%B*H/(1+%B*H))**2+
             %GT(%B,0)*(%EXP(-%B*H)*(-%B*H-1)+1)**2)
   : VA=%VA*VA/%A**4 : VB=%VA*VB/%B**4
   : %V=%CU(VA) : %W=%CU(VB)
   : C1=%GE(H,0)*(%GT(%A,0)*%EXP(%A*H)*(%EXP(%A*H)*(%A*H-1)+1)+
            %LT(%A,0)*(%LOG(1-%A*H)+%A*H/(1-%A*H))/(1-%A*H))
   : C2=-%LT(H,0)*(%LT(%B,0)*(%LOG(1+%B*H)-%B*H/(1+%B*H))/(1+%B*H)+
            %GT(%B,0)*%EXP(-%B*H)*(%EXP(-%B*H)*(-%B*H-1)+1))
   : C1=C1*%VA/%A**2 : C2=C2*%VA/%B**2  $EXTRACT %VC
$C %V,%W ARE DIAG(U);C1,C2 COLUMNS OF C;
$C  %VC THE VARIANCE OF BETA GIVEN ALPHA, USING NOTATION IN TEXT
$OUTPUT 3 $PRINT %V %W %VC $LOOK DOSE C1 C2 $OUTPUT $ENDMAC

$MACRO DER
$C CALCULATES DERIVATIVES OF H WRT (B,A) FOR VARIANCE OF FITTED VALUES
$CALC H1=1 :H2=DOSE :H=%LP :HA=%A*%LP :HB=%B*%LP
    :HA1=%GE(H,0)*(%GT(%A,0)*(%EXP(HA)*(HA-1)+1)
             +%LT(%A,0)*(%LOG(1-HA)+HA/(1-HA)))
    :HA2=-%LT(H,0)*(%LT(%B,0)*(%LOG(1+HB)-HB/(1+HB))
             +%GT(%B,0)*(%EXP(-HB)*(-HB-1)+1))

    :HA1=HA1/%A**2 :HA2=HA2/%B**2
```

```
    :HBB=%GE(H,0)*(%GT(%A,0)*%EXP(HA)+%LT(%A,0)/(1-HA)+%EQ(%A,0))
      +%LT(H,0)*(%LT(%B,0)/(1+HB)+%GT(%B,0)*%EXP(-HB)+%EQ(%B,0))
    :HBX=HBB*DOSE
$OUTPUT 9 80 $LOOK H1 H2 HBB HBX HA1 HA2 $OUTPUT $ENDMAC

$UNITS 8 $DATA DOSE Y N $DINPUT 4$
$C FIT THE STANDARD LOGISTIC MODEL
$YVAR Y $ERROR B N $FIT DOSE $
      SCALED
 CYCLE  DEVIANCE     DF
   4    11.23        6

$DISP E V $CALC P=Y/N :F=%FV/N $PLOT F P DOSE $
     ESTIMATE      S.E.      PARAMETER
   1  -60.72       5.180      %GM
   2   34.27       2.912      DOSE
  SCALE PARAMETER TAKEN AS    1.000

(CO)VARIANCE MATRIX
 1   26.83
 2  -15.08     8.478
       1         2
  SCALE PARAMETER TAKEN AS    1.000

 1.05      *
 1.00      *                          P   2
 0.950     *                           F
 0.900     *                       2
 0.850     *                    P
 0.800     *                    F
 0.750     *
 0.700     *
 0.650     *
 0.600     *                 F
 0.550     *
 0.500     *                 P
 0.450     *
 0.400     *
 0.350     *            F
 0.300     *            P
 0.250     *
 0.200     *       P
 0.150     *       F
 0.100     * P
 0.500E-01 *  F
 0.000E+00 *
 .........*.........*.........*.........*.........*.........*
        1.67      1.72      1.77      1.82      1.88      1.92

$OWN M1 M2 M3 M4 $SCALE 1 $CALC %E=.00000001$
----- CURRENT DISPLAY INHIBITED
$C STARTING VALUE IS LINEAR PREDICTOR FROM LOGISTIC FIT
$CALC %LP=%LOG(F/(1-F)) : H=%EQ(F,0)+%EQ(F,1)
   : %LP=50*%EQ(F,1)-50*%EQ(F,0)+%LP*%EQ(H,0)

$C SET SCALE PARAMETERS A1 AND A2; PERFORM INTERACTIVE FIT
$CALC %A=0.16 : %B=-0.53            !
$USE INT$                           !

      SCALED
 CYCLE  DEVIANCE     DF
   3    2.742        6
```

```
     ESTIMATE      S.E.      PARAMETER
  1  -69.94       7.186      %GM
  2   39.25       3.991      DOSE
  SCALE PARAMETER TAKEN AS    1.000

 1.05        *
 1.00        *                                         2     2
 0.950        *                                   G
 0.900        *                                   P
 0.850        *                              P
 0.800        *                              G
 0.750        *
 0.700        *
 0.650        *
 0.600        *
 0.550        *
 0.500        *                        2
 0.450        *
 0.400        *
 0.350        *
 0.300        *                  2
 0.250        *
 0.200        *           2
 0.150        *
 0.100        * 2
 0.500E-01 *
 0.000E+00 *
 .........*.........*.........*.........*.........*.........*
         1.67          1.72          1.77         1.82         1.88         1.92


 $C CALCULATE VARIANCE COMPONENTS
 $USE VCOV $USE DER$
```

I wish to thank David Andrews for his constructive comments and helpful discussions and the referees for their useful suggestions. This research was funded in part by the Natural Sciences and Engineering Research Council of Canada, Grant A8186.

REFERENCES

BAKER, R.J. and NELDER, J.A. (1978). The GLIM Manual-Release 3. Distributed by Numerical Algorithms Group, Oxford.

BLISS, C.I. (1935). The calculation of the dosage-mortality curve. Ann. Appl. Biol. 22, 134-167.

COX, D.R. (1970). The Analysis of Binary Data. Chapman and Hall, London.

NELDER, J.A. and WEDDERBURN, R.W.M. (1972). Generalized Linear Models. JRSS(A) 135, 370-384.

STUKEL, T. (1983). Generalized Logistic Models. Ph.D. thesis, University of Toronto.

Perturbing a Sparse Contingency Table

by JOE WHITTAKER

Department of Mathematics
University of Lancaster, England

SUMMARY

The robustness of contingency table analysis is examined by perturbing the data in the sparse regions of the table. The effect of the perturbation is assessed by the change to the additive elements of the conditional equi-probability models. By way of example it is concluded that the graphical models are substantially less robust than the corresponding two-way interaction models.

Keywords: Additive Elements; Contingency Table; Graphical Model; Log-linear Model; Perturbation; Sparse data; Two-way Interaction Model.

1. INTRODUCTION

No statistical analysis should hang on the outcome of deleting a single observation from the sample. In regression analysis this has generated a literature devoted to methods for identifying outliers and measures of influence recently summarised in the text of Cook and Weisberg (1982). While, most of these techniques do not readily extend to contingency table analysis, perhaps because of the inherent grouping in the data and perhaps because of the multi-dimensional nature of the concept of influence, perturbation is a technique that does.

This paper gives an example of the effect of deleting one or two of the potentially most sensitive observations from the table. A priori considerations suggest that the analysis will be most sensitive to perturbing an observation in a cell with relatively few entries. Rather than perturb all cells we restrict attention to just these. Though there are many ways to measure influence some will not work satisfactorily. For example, suppose the analysis bases model selection by inspection of the parameters of the saturated model. In one well-known parameterisation of the log-linear model adjusting the last cell in the table affects only the value of the last parameter, while adjusting the first affects them all. Our measure of influence is closely bound up with the notions of graphical models and the additive elements of decomposable models.

The application of the notion of zero partial association to model selection by Wermuth (1976) with its intimate relation to the graphical ideas of Darroch, Lauritzen and Speed (1980) have had a substantial impact in the analysis of contingency tables. A very readable account is given by Edwards and Kreiner (1983). Restricting attention to the subset of graphical models goes some way to simplifying the task of the analyst; however the graphical class is by no means small. Consider the data from Stouffer and Toby (1951) on attitude.

Table 1. Role conflict response of 216 respondents.

C	+		-			
D	+	-	+	-	A	B
	42	23	6	25	+	+
	6	24	7	38	+	-
	1	4	1	6	-	+
	2	9	2	20	-	-

Each respondent was presented with four situations: A, B, C and D. The response to each was either universalistic (+) or particularistic (-). There are 4choose2=6 conditional independences of the form $A \perp B \mid C,D$ in which the pair of factors A and B are conditionally independent of the remaining factors, for short, the 'rest'. Each graphical model is generated by including a subset of these independences.

As with many other model selection procedures the deviances of a variety of models have to be computed. Whittaker (1982) adapted the simultaneous test procedure of Aitken (1979) to select a graphical model which appeared to require computing the deviances of 6!=720 models. In computing these Whittaker (1982) noted that the deviances are additive over certain models.

Example. Consider the independence graphs

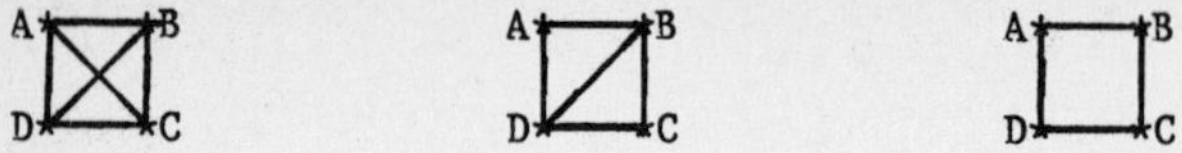

The second graphical model is reducible and satisfies the addition laws

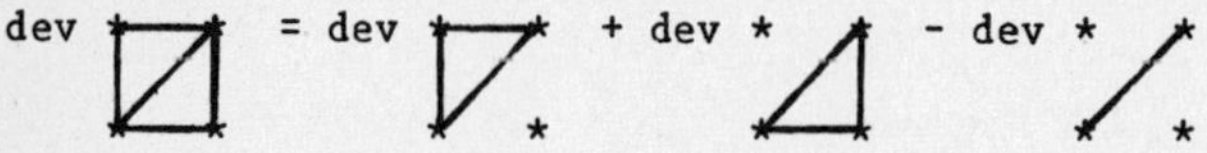
dev
= dev
+ dev
- dev

and

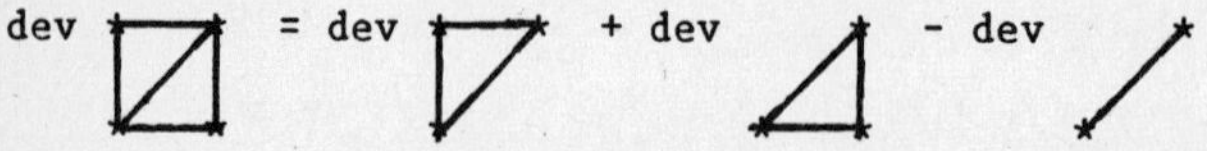
dev
= dev
+ dev
- dev

While the first expression reduces the calculation of the deviances of all graphical models to calculating those of 14 elementary graphical models, the second expression allows a more systematic exploitation of additivity and was used by Whittaker (1984b). In model terms it reads

$$\text{dev}(A.B.D+B.C.D) = \text{dev}(A.B.D) + \text{dev}(B.C.D) - \text{dev}(B.D)$$

rather than

$$\text{dev}(A.B.D+B.C.D) = \text{dev}(C+A.B.D) + \text{dev}(A+B.C.D) - \text{dev}(A+C+B.D).$$

It reduces the calculation of the deviances of all decomposable models to the calculation of 16 models. These models for example A or A.B.C, have just a single generator in their generating class and can be described as models of conditional equi-probability. Thus dev(A.B) is a test for the equi-probability of the 4 levels of the combined C,D factor at each of the four levels of the combined A,B factor.

Having reduced the deviances of the decomposable models to the deviances of the conditional equi-probability models the next step is to transform to additive elements. They are a generalisation of the regression elements of Newton and Spurrell (1967), see Whittaker (1984a) for a discussion and were used in contingency tables by Whittaker (1984b). They give a near one to one representation of a 2^k contingency table in which

- the primary elements explicitly give the deviances for tests of the equi-probability of one factor given the 'rest';

- the secondary elements give the deviances for pairwise conditional independence given the 'rest'; and
- the third and higher order elements are differences between tests for conditional independence based on varying the conditioning set.

The transformation is particularly simple to code in GLIM, Whittaker (1985).

This suggests that one way to measure the influence of a perturbation is by its effect on the additive elements of the table. Though multi-dimensional it is a concise representation of the dependence relations manifest in the table. Because no iterative fitting or matrix inversion is required to compute the elements this is a viable technique to use for large multi-way tables and for simulation experiments.

2. ADDITIVE ELEMENTS of the CONDITIONAL EQUI-PROBABILITY MODELS

The deviances and degrees of freedom of each conditional equi-probability model for the data of Table 1. are computed from expressions such as

$$dev(A.B) = 2 \sum n(i,j,k,l) \log \{ n(i,j,k,l).K.L \;/\; n(i,j,+,+) \}$$

where n gives the number of observations in each cell.

Table 2. Deviances and degrees of freedom of equi-probability models.

model	dev	df	model	dev	df
A.B.C.D	0.0	0	A.B.C	81.0	8
A.B.D	38.4	8	A.B	100.4	12
A.C.D	43.8	8	A.C	107.2	12
A.D	71.9	12	A	113.2	14
B.C.D	99.2	8	B.C	175.6	12
B.D	135.3	12	B	191.5	14
C.D	134.7	12	C	191.4	14
D	159.6	14	1	191.5	15

A GLIM macro for the computation of the elements is given in Whittaker (1985). The transformation is based on noting that the additive elements are just the parameter estimates of the saturated model for a 2^k complete factorial design in which identifiabilty is ensured by using the convention incorporated in GLIM that the parameter corresponding to the first factor level is set to zero. The result is the following table:

Table 3. Additive Elements of the conditional equi-probability models.

order	element	transformed dev	df
0	:ABCD	0	0
1	A:BCD	99.15	8
	B:ACD	43.78	8
	C:ABD	38.40	8
	D:ABC	81.00	8
2	AB:CD	-8.22	-4
	AC:BD	-2.29	-4
	BC:AD	-10.29	-4
	AD:BC	-4.56	-4
	BD:AC	-17.53	-4
	CD:AB	-19.00	-4
3	ABC:D	-0.92	2
	ABD:C	-2.25	2
	ACD:B	-1.16	2
	BCD:A	-3.19	2
4	ABCD:	-1.38	-1

The primary elements are tests for equi-probability and the secondary elements are tests for conditional independence. Since the latter are of more interest we suppress the primary elements in the next figure.

Figure 1. Additive Elements Plot.

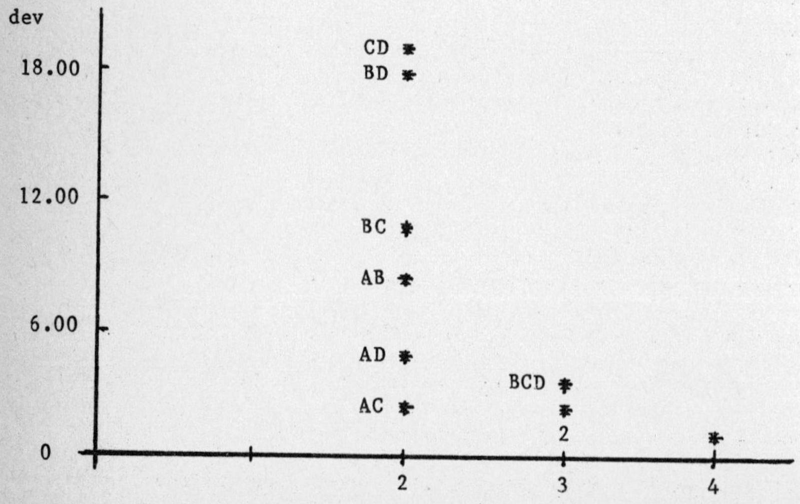

The essential result for interpreting this plot is that the second order element, such as G(AB:CD), is the negative of the maximised log-likelihood ratio test for the conditional independence of factors A and B given the remaining factors C and D. As such it has an asymptotic chi-squared distribution on the null hypothesis with degrees of freedom given by the negative of the corresponding transformed degree of freedom. The elements plot in Figure 1. reveals relatively strong dependences between factors C and D and between B and D and weaker dependences between A and D and between A and C. The 5% significance level of chi-square(4) is 7.78 so that a crude selection procedure is to choose the graphical model

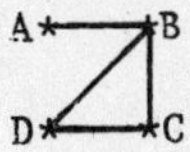

This model was in fact the best of those selected by the simultaneous test procedure of Whittaker (1982).

3. MARGINAL TABLES

The graphical model chosen for this data, A.B+B.C.D, makes the following statements about the four marginal 3-way tables:

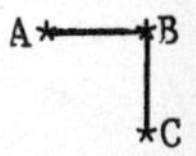

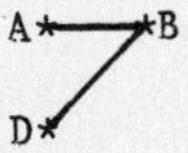

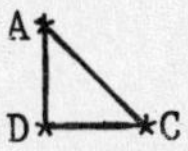

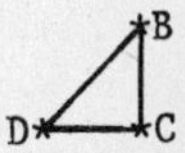

Additivity allows us to check if these are consistent with the data. We only need to check for independences because the third and fourth order elements in Table 2. are all of the same sign as the secondary elements. Hence any test for conditional independence in a marginal table will have a larger value than in the full table. For example additivity implies that the test for $A \perp B \mid C$ is

$$G(AB:C) = G(AB:CD) + G(ABD:C) = - 8.22 - 2.25 = -10.47$$

on -4+2=-2 df; it has a higher P-value than the test for $A \perp B \mid C,D$. Consider the test for $A \perp C \mid B$, the value of

G(AC:B) = G(AC:BD) + G(ACD:B) = -2.29 - 1.16 = -3.45 on 2df

is in accord with the independence predicted from the graph. A similar calculation shows that G(AD:B)=-5.72 as compared to the 5% point of chi-square(2) of 5.99; and that G(AD:C)=6.81. The only slight variation from predicted behaviour is that G(AC:D)=-3.21 in the third marginal table.

No two-way marginal independences are predicted from the graphical model nor are obtained from additivity calculations. For example,

G(AC:) = -2.29 - .92 - 1.16 - 1.38 = -5.75 on -4+2+2-1=-1 df

In passing we note that the deviances for tests of conditional independence and marginal independence for each pair of variables was given in the deviance difference matrix of Whittaker (1982, Table 3.). The secondary elements of Table 3. agree with those as do the test statistics for marginal independence, such as G(AC:), extracted by additivity.

4. PERTURBATIONS and ADJUSTMENTS for SPARSE TABLES

We consider perturbing the data of Table 1. by replacing the two entries with a single count by a zero count. Recomputation of the additive elements will indicate if the analysis is robust to this perturbation.

Table 4. Perturbed data.

C	+		-			
D	+	-	+	-	A	B
	42	23	6	25	+	+
	6	24	7	38	+	-
	0	4	0	6	-	+
	2	9	2	20	-	-

When a table has zero entries in marginal tables corresponding to a generator in the generating class then sometimes an iterative method of computing the maximum likelihood estimates may not converge and some adjustment to the degrees of freedom is necessary. See Brown and Fuchs (1983), Aston and Wilson (1984).

Consider the models of conditional equi-probability: the deviance of A.B.C is a combined test that the levels of D are equi-probable at each combination of the levels of A, B and C. Thus there are 8 tests with outcomes (42,23), (6,25), (6,24), (7,38), (0,4), (0,6), (2,9) and (2,20). No adjustment is necessary. However the corresponding test based on A.B.D contains the pair (0,0) which gives no information about equi-probability. The deviance is thus combined from 7 rather than 8 informative tests and consequently must be adjusted by subtracting 1 df. This effects many of the other tests. No other overall test for conditional equiprobability has any non-informative comparisons, so the deviances are recomputed for this perturbed table and transformed to additive elements.

Table 5. Additive Elements of the Perturbed data

order	element	transformed dev	df
0	:ABCD	0	0
1	A:BCD	111.62	8
	B:ACD	48.64	8
	C:ABD	38.40	7
	D:ABC	88.97	8
2	AB:CD	-13.77	-4
	AC:BD	-0.74	-3
	BC:AD	-10.29	-3
	AD:BC	-10.81	-4
	BD:AC	-19.73	-4
	CD:AB	-18.94	-3
3	ABC:D	-1.81	1
	ABD:C	0.84	2
	ACD:B	-2.59	1
	BCD:A	-3.07	1
4	ABCD:	-0.79	0

All tests for conditional independence in which C is part of the conditioning set have the same number of degrees of freedom as in Table 3., but those for which C is one of the arguments before the colon have lost one. Exactly the same has happened to the higher order elements.

The changes in the observed values of the elements are fairly disquieting.

Only the element G(C:ABD) and G(BC:AD) remain unaltered. Some move quite substantially, for instance, G(AB:CD) from -8.22 to -13.77. The change in G(AD:BC) from -4.56 to -10.81 clearly changes the model selected. The change is not all in one direction, for instance G(AC:BD) moves from -2.29 to -.74 while G(ABD:C) even changes sign. Elements in marginal tables are correspondingly effected.

Though it is hard to discern any overall pattern of change it appears that this contingency table and its additive elements are not robust to the perturbation. This conclusion extends to all decomposable (and almost all graphical) models as can be directly generated from the conditional equi-probability models. Satisfaction with the fitted model as a true approximation to the data is accordingly diminished.

5. ALL TWO-WAY INTERACTION MODELS

The analysis in the previous section demonstrates that the decomposable models are sensitive to the perturbation of the table obtained by deleting two entries. Consider replacing the equiprobability models by the corresponding two way interaction model. The reason for such a replacement is that the sufficient statistics for any two-way interaction model are never more than two-dimensional. Hence it is expected that such models are more robust to sparsity.

The correspondence is A.B.C.D goes to A.B+A.C+A.D+B.C+B.D+C.D, A.B.C goes to A.B+A.C+B.C, A.B goes to A.B, and so on. Deviance additivity goes through for these models as well. The deviances transformed to additive elements are:

Table 6. Additive Elements of two-way interaction models.

order	element	transformed dev original	perturbed	df
0	:ABCD	8.49	11.57	5
1	A:BCD	100.30	105.17	4
	B:ACD	37.95	37.71	4
	C:ABD	34.11	33.28	4
	D:ABC	77.53	78.53	4
2	AB:CD	-7.31	-7.97	-1
	AC:BD	-0.61	-0.73	-1
	BC:AD	-6.03	-5.81	-1
	AD:BC	-4.25	-5.50	-1
	BD:AC	-10.79	-9.94	-1
	CD:AB	-15.63	-14.94	-1
3	ABC:D	-1.04	-1.17	0
	ABD:C	-3.40	-3.83	0
	ACD:B	-1.21	-1.46	0
	BCD:A	-7.03	-6.42	0
4	ABCD:	-2.33	-2.57	0

Apart from the residual term the original and the perturbed elements are almost identical. The one difference is that the element of order zero, the residual, which is the goodness of fit test for the model of all two way interactions, dev(A.B+A.C+A.D+B.C+B.D+C.D) on 5 df, which would be rejected on the perturbed data set but not from the original data.

6. DISCUSSION

The example pursued by this paper is an attempt to come to grips with the problem of assessing robustness. Perturbation is one way to examine the sensitivity of the data and of the proposed model to slight changes in the table. Though it seems reasonable to perturb in sparse regions of the table we make no proposal to systematise the technique which would clearly be necessary before embarking on more rigorous simulation studies.

Concentrating on the additive elements focusses attention on the conditional independence relationships manifest by the factors cross classifying the table. Comparison of the perturbation effect on the elements clearly show that the graphical models are less robust than their two way interaction counterparts.

6. REFERENCES

Aston, C.E. and Wilson, S.R. (1984) Comment on M.B.Brown and C.Fuchs. Computational Statistics and Data Analysis 2, 71-77.

Brown, M.B. and Fuchs, C. (1983) On maximum likelihood estimation of sparse contingency tables. Computational Statistics and Data Analysis 1, 3-15.

Cook, R.D. and Weisberg, S. (1982). Residuals and influence in regression. N.Y. and London, Chapman and Hall.

Darroch, J., Lauritzen, S. and Speed, T. (1980). Markov fields and log linear interaction models for contingency tables. Ann. Stat. 8. 522-539.

Edwards, D. and Kreiner,S. (1983). The analysis of contingency tables by graphical models. Biometrika. 70, 3, 553-565.

Newton, R.G. and Spurrell, D.J. (1967) A development of multiple regression for the analysis of routine data. Appl. Stat. 16, 51-64.

Whittaker, J.(1982) Glim syntax and simultaneous tests for graphical log-linear models. In Gilchrist R. (Ed). GLIM 82. Lecture notes in Statistics, Vol.14, 98-108. Springer Verlag.

Whittaker, J.(1984a) Model interpretation from the additive elements of the likelihood function. Applied Statistics, 1984, 33,1,52-64.

Whittaker, J.(1984b) Fitting all possible decomposable and graphical models to multiway contingency tables. In Havranek,T. et al.(Ed) Compstat 1984. 401-406. Physica-Verlag. Vienna.

Whittaker, J.(1985) Transforming to the additive elements of the deviance. (to appear) Glim Newsletter. 1985, 10.

Wermuth, N. (1976). Analogies between multiplicative models in contingency tables and covariance selection. Biometrics, 32, 95-108.

Wermuth, N. (1976). Model search among multiplicative models. Biometrics, 32, 253-264.